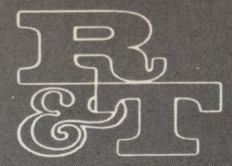

ROAD & TRACK
ON
COBRA, SHELBY & FORD GT40
1962-1983

Reprinted From
Road & Track Magazine

ISBN 0 946489 36 X

Published By
Brooklands Books with permission of Road & Track
Printed in Hong Kong

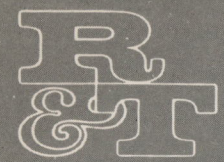

Titles in this series:

Road & Track on Alfa Romeo 1977-1984
Road & Track on Austin Healey 1953-1970
Road & Track on BMW Cars 1975-1978
Road & Track on BMW Cars 1979-1983
Road & Track on Cobra, Shelby & Ford GT40 1962-1983
Road & Track on Corvette 1953-1967
Road & Track on Corvette 1968-1982
Road & Track on Datsun Z 1970-1983
Road & Track on Ferrari 1950-1968
Road & Track on Ferrari 1968-1974
Road & Track on Ferrari 1975-1981
Road & Track on Fiat Sports Cars 1968-1981
Road & Track on Jaguar 1968-1974
Road & Track on Jaguar 1974-1982
Road & Track on Lamborghini 1964-1982
Road & Track on Lotus 1972-1983
Road & Track on Maserati 1952-1974
Road & Track on Maserati 1975-1983
Road & Track on Mercedes Sports & GT Cars 1970-1980
Road & Track on Porsche 1968-1971
Road & Track on Porsche 1972-1975
Road & Track on Porsche 1975-1978
Road & Track on Porsche 1979-1982
Road on Track on Triumph Sports Cars 1974-1982

Titles in preparation will cover:
Aston Martin, Bentley, Lagonda, MG & Rolls Royce
plus further titles on Alfa Romeo, BMW, Jaguar & Lotus

Distributed By

Road & Track
1499 Monrovia,
Newport Beach,
California 92651, U.S.A.

Brooklands Book Distribution Ltd.,
Holmerise, Seven Hills Road,
Cobham, Surrey KT11 1ES,
England

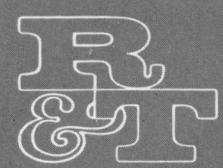

Contents

5	AC-Ford Cobra Road Test	*Road & Track*	Sept.	1962
10	Corvette vs. Cobra: The Battle for Supremacy	*Road & Track*	June	1963
13	AC Cobra Road Test	*Road & Track*	June	1964
17	Ford GT	*Road & Track*	June	1964
18	Ford GT	*Road & Track*	July	1964
21	A Look at the Daytona Winner Ford GT40	*Road & Track*	May	1965
25	Mustang GT350 Road Test	*Road & Track*	May	1965
32	Ford GT 40	*Road & Track*	June	1965
34	Cobra Wins Le Championnat Des Constructeurs	*Road & Track*	June	1965
36	Ford GT Mark II	*Road & Track*	April	1966
37	The Challenge to Ferrari— Porsche Carrera/Ford GT Mark II	*Road & Track*	May	1966
42	Portrait of the Le Mans Winner Technical Analysis	*Road & Track*	Oct.	1966
50	Shelby American Mustangs for 1967	*Road & Track*	Jan.	1967
54	Shelby GT 500 Road Test	*Road & Track*	Feb.	1967
58	Shelby American: 1968	*Road & Track*	Jan.	1968
60	Cobra To End All Cobras	*Road & Track*	Feb.	1968
62	Two Shelby GT350s Road Test	*Road & Track*	June	1968
66	24 Heures Du Mans 1969 — Ford's 4th Straight	*Road & Track*	Sept.	1969
73	Cobra 427 Classic Road Test	*Road & Track*	July	1974
78	1964 FIA Cobra Roadster Salon	*Road & Track*	Dec.	1979
84	GT40 Reborn	*Road & Track*	Aug.	1981
86	Cobra Replicas	*Road & Track*	Aug.	1983
91	Carroll Shelby — Profile	*Road & Track*	Aug.	1983

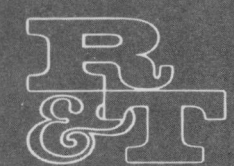

We are frequently asked for copies of out of print Road Tests and other articles that have appeared in Road & Track. To satisfy this need we are producing a series of books that will include, as nearly as possible, all the important information on one make or subject for a given period.

It is our hope that these collections of articles will give an overview that will be of value to historians, restorers and potential buyers, as well as to present owners of these automobiles.

Copyright © CBS Inc.

AC-FORD COBRA

An Ace from AC, assisted by a V-8 from Ford, is a 150-mph car built for production racing

THE DIRECT APPROACH in the production of a high-performance car is to combine a small, light chassis and body, and a big, powerful engine. It may not be subtle, but it is sure-fire, and no one could be better acquainted with this fact than Carroll Shelby. The Pride of Texas has spent a lot of years charging about in all kinds of automobiles and has a fine appreciation of what will, in practical terms, get a person down the road in a hurry. Hence, when Shelby retired from competition driving and began making plans to embark on a venture into automobile manufacturing, it was inevitable that the direct, and effective, approach would be employed.

After due deliberation and prolonged negotiation (the details of which we do not pretend to know) agreements were reached with AC, of England, to make the basic automobile for Shelby, and Ford, of America, to supply engines.

The Cobra's chassis is a very close development of the latest AC "Ace." The frame follows almost exactly the standard AC pattern, having a pair of round-section, tubular main-members running parallel between the box structures, to which the front and rear suspension links and springs are mounted. However, the tubes in the Cobra frame have a slightly greater wall thickness and there is some extra cross bracing.

All 4 wheels are independently suspended. The suspension consists of A-arms leading from the box structures on the frame to the lower ends of the "uprights" at the wheels and transverse leaf springs (one at each end of the chassis) completing the parallelogram at the upper ends of the uprights.

At Shelby's request, the A-arms and leaf-springs have been lengthened to move the wheels out and widen the tread. The springing has been stiffened by the addition of a "helper" leaf above the main leaf. Also, the rear wheels are given a considerable negative camber to reduce the oversteering effects of the car's rearward weight bias. The suspension system is lacking in some respects, as its parallelogram action tilts the wheels over when the chassis rolls, but it is all-independent, as all too few touring chassis are, and it gives good results.

The AC's usual 16-in. wire wheels are replaced by 15-in. triple-laced, wide-rim wheels. These carry a variety of tires —depending on customer preference—but most of the cars will be Goodyear-shod. Dunlop racing tires will be catalogued as an option but Ol' Shel has said that anything a buyer wants, he will get. We would offer one bit of advice to prospective purchasers; select a tire that gives plenty of tread against the ground; with the Cobra's power/weight ratio, you will need the traction.

Traction will also be needed to get the maximum benefits from the Cobra's braking system. Girling disc brakes are standard, using discs 12 in. in diameter and, although there is no power-booster, the pedal pressure requirements are not too high. On the prototype, the rear brakes were mounted inboard, one on each side of the drive casing, but an outboard mounting will probably be standard on the series-produced cars.

Power for the Cobra is one of Ford's new lightweight V-8s.

It is the largest of the Fairlane-series engines, having a displacement of 260 cu in. and, in single-carburetor form, pushes out an "advertised" 260 bhp. Our test car had the "street" engine, which is virtually stock Ford but equipped with solid valve lifters and a camshaft of non-standard but unspecified timing. We were somewhat surprised to find that it would idle with only a trace of lumpiness, pull strongly at almost any speed, and buzz past the 5800 rpm power peak to 7200 rpm before it began to sound distressed. Also surprisingly, the power did not appear to fall off much even at 7000 rpm—1200 rpm over the point of maximum power.

For the buyer who wants to go racing Shelby has something special under development: a highly tuned competition version of the standard engine with a "top-end-grind" camshaft, higher compression and 4 double-throat Weber carburetors. These are side-draft pattern carburetors, and are mounted on long ram tubes that criss-cross over the top of the engine. In tuned form, the Cobra/Ford engine will produce about 325 bhp, and the added power should do really impressive things for a car that is blindingly fast on 260 bhp.

Special drive components are used all the way through in the Cobra. The transmission is from Borg-Warner (which makes these 4-speed units for Ford) and, although the prototype car had rather wide ratios, Borg-Warner—and Ford—have decided to invest in the tooling for close ratio gearing. Such gearing would eliminate the "gap" that now exists between 3rd and 4th, and the top-end acceleration would be substantially improved. With the present gear staging, the shift into 4th entails a drop in engine speed that would be a serious handicap in competition.

Final drive components—gears, axles, U-joints, etc.—are heavier than those used in the standard AC. A Salisbury drive-gear assembly, of the type used by Jaguar, fits into a housing that is part of the frame (per long-time AC practice) and drives the wheels through Spicer shafts and U-joints. Both shafts and joints have been selected for the strength needed to transmit the power from the Ford engine.

The Ford Fairlane V-8 fits the AC engine compartment without the proverbial shoehorn.

Bodywork on the Cobra is almost, but not quite, the same as on the latest AC Ace roadsters. The only change worth mentioning is the flared valance over each wheel well. These are required to provide clearance for the Cobra's large-section tires and widened tread. Shelby is considering making up a few Cobras with AC's Aceca (coupe) body and these would have a top speed even higher than the roadster's.

Within, the Cobra has even more racing-car flavor than is apparent from without. The cockpit (that term exactly describes the passenger area in the Cobra) is quite snug, and the leather-covered bucket seats are real hip-huggers, with lateral support that extends almost from shoulder to knee. Instrumentation is more elaborate than in most racing cars, consisting of speedo, tach, oil-pressure and temperature and water temperature, with a clock, an ammeter and fuel gauge tossed in for good measure (ouch!). The speedo had been removed from the prototype we tested, as the instrument supplied

ROAD TEST
AC FORD COBRA

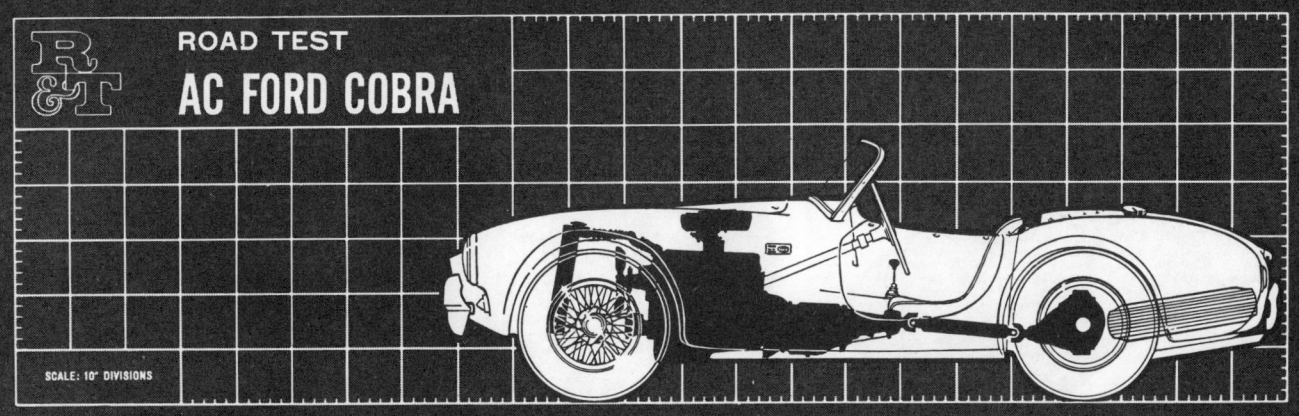

SCALE: 10" DIVISIONS

DIMENSIONS
Wheelbase, in	90.0
Tread, f and r	51.5/52.5
Over-all length, in	151.5
width	61.0
height	49.0
equivalent vol, cu ft	262
Frontal area, sq ft	16.6
Ground clearance, in	7.0
Steering ratio, o/a	n.a.
turns, lock to lock	2.0
turning circle, ft	34
Hip room, front	2 x 16.5
Hip room, rear	n.a.
Pedal to seat back, max	40.0
Floor to ground	10.5

CALCULATED DATA
Lb/hp (test wt)	9.6
Cu ft/ ton mile	175.2
Mph/1000 rpm (4th)	21.8
Engine revs/mile	2745
Piston travel, ft/mile	1315
Rpm @ 2500 ft/min	5230
equivalent mph	114.3
R&T wear index	36.1

SPECIFICATIONS
List price	$5995
Curb weight, lb	2020
Test weight	2355
distribution, %	48/52
Tire size	6.50/6.70-15
Brake swept area (est)	580
Engine type	V-8, ohv
Bore & stroke	3.80 x 2.87
Displacement, cc	4261
cu in	260
Compression ratio	9.2
Bhp @ rpm	260 @ 5800
equivalent mph	127
Torque, lb-ft	269 @ 4500
equivalent mph	98

GEAR RATIOS
4th (1.00)	3.54
3rd (1.41)	4.99
2nd (1.78)	6.30
1st (2.36)	8.36

SPEEDOMETER ERROR
30 mph	actual, n.a.
60 mph	n.a.

PERFORMANCE
Top speed (7000), mph	153
best timed run	n.a.
3rd (7200)	112
2nd (7200)	89
1st (7200)	67

FUEL CONSUMPTION
Normal range, mpg	n.a.

ACCELERATION
0-30 mph, sec	1.8
0-40	2.5
0-50	3.3
0-60	4.2
0-70	5.4
0-80	6.8
0-100	10.8
Standing ¼ mile	13.8
speed at end	112

TAPLEY DATA
4th, lb/ton @ mph	off scale
3rd	off scale
2nd	off scale
Total drag at 60 mph, lb	115

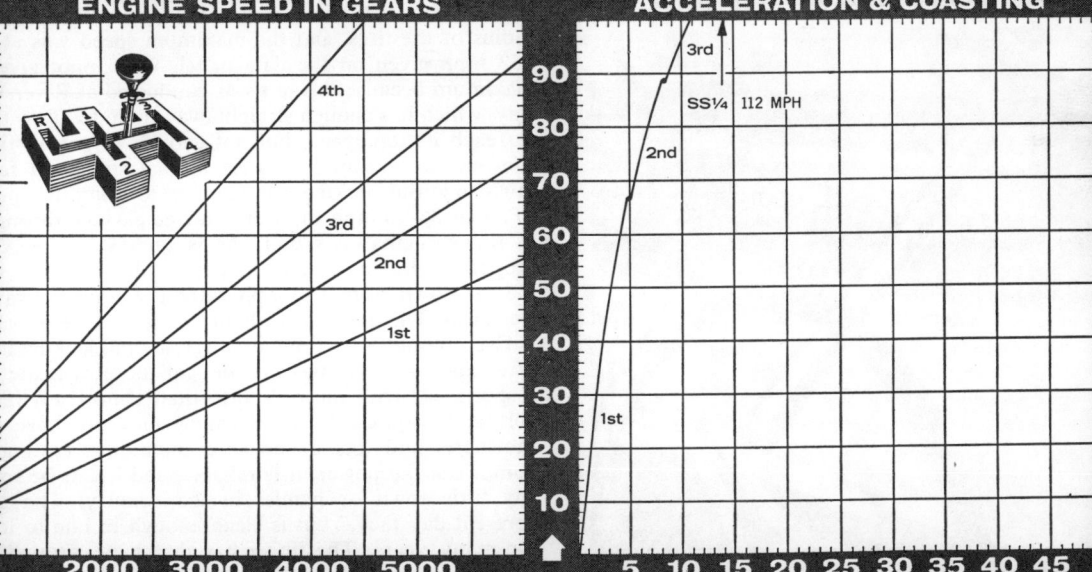

A full complement of instruments and a racing steering wheel add a touch of glamour.

Bucket seats are a bit too upright, but still are very comfortable.

After the spare tire goes in, there isn't too much space left. But it is usable.

AC FORD COBRA

by AC had a scale that reached only the 120-mph mark.

Control positioning isn't bad—if you are less than 6 feet tall. The wheel is a bit close for a driver with long arms and the pedals (clutch and brake) are manipulated with one's knees drawn well up. Shelby tells us that he plans adjustments to give more room and we certainly hope this is done. Our tall-ish test driver had his work complicated slightly by the lack of clearance, which resulted in his knees banging against the under edge of the instrument panel with each application of the clutch or brake. Part of this was due to the positioning of the roll-bar in the car; it was placed so that the seat could not be moved back far enough.

In retrospect, we can see that the close-up position of the steering wheel on the prototype may have had its advantages. The Cobra had very quick steering, and what seemed like a lot of caster in the front wheels, and considerable effort is required at the wheel rim to force the car into corners at anything faster than a touring pace. With the wheel in close, it was easier to apply the needed muscle.

In a demonstration at Riverside Raceway, Carroll Shelby shows photographer Bill Motta 6000 in 4th gear.

Despite the less-than-perfect control positioning (which should not be true of the series-produced cars) the Cobra's handling was good. With so much power on tap, the inept or inexperienced could get into considerable trouble, but a middlin'-good driver can certainly get the car around a race course in a hurry. One facet of the handling that made us feel a trifle wary at first was the extreme angle (relative to its true line of travel) the car assumes when drifting. There is some oversteer, and when the Cobra is shoved into a turn with brio, the rear wheels creep right out. Treated with any finesse at all, the Cobra will hold its tail-out attitude without trying to spin, but a clumsy throttle foot could give you a thrill.

Insofar as sheer speed is concerned, the Cobra offers more than almost any sports/touring car in the world, and more than any at near its price. Its acceleration, even with the "small" engine, is equal to the best efforts of drag-strip-tuned Corvettes, and it does the job without the benefit of stump-yanker gearing. No special talent is required to get under the 14-sec mark for a standing-start ¼-mile; bang down the throttle-pedal, simultaneously drop in the clutch, and catch the next higher gear each time you reach 7200 rpm. If you persist, the car will accelerate until the tachometer shows 7000 rpm in 4th, which is, without making any allowance for tire expansion, 153 mph. This speed was reached with the car in touring trim, and we were at first reluctant to believe it ourselves. However, we checked and double-checked the accuracy of the tachometer, the gear ratio and the rolling-radius of the tires, and the maximum speed was at least the 153 mph given on the data panel. We cannot give a timed maximum because this test was conducted at Riverside Raceway, which has enough straightaway room to allow the Cobra to reach its top speed, but not enough to permit timing the car over a measured distance at that speed. In fact, some fairly vigorous braking was required to bring the speed down to about 85-90 mph for a drama-free passage through turn 9 at the straightaway's end. As is understandable, we were properly delighted to find that the Cobra's brakes will yank the car down from 150 mph without a trace of wavering or weakness. The brakes are about the best we've ever tried.

Even though something of a hybrid, and lacking in the sort of engine-room niceties that delight the purist, the Cobra is a sports car with more "sport" than almost anything available at any price. Its Ford engine may not have overhead camshafts and lots of polished aluminum castings, but it pumps out the power, it is reliable, and it can be serviced in any little town or hamlet in the country. The styling is "present-day racy," but is clean enough in line to look good for many years. The finish has that hand-wrought appearance that cannot be machine-made and the Cobra looks every bit as good as its price tag. Indeed, we cannot think where more all-around performance can be purchased at the same dollar outlay.

MEN WHO KNOW FINE CARS, APPRECIATE THE COBRA!

It's a rare **COBRA** owner who can't instantly recognize the characteristic red lights and apologetic grin that seem to appear occasionally, without reason, in his rear view mirror. No ticket involved, just admiration and interest!

Perhaps it's the way a **COBRA** glides effortlessly through traffic, its four wheel independent suspension just rippling over the road. Or maybe it's the sound... the matchless cadence of power with an American accent — Ford 289 High Performance V-8! Whatever it is, there's just *something* about a **COBRA** that demands more than a casual glance, and our white helmeted friends in blue are no exception!

Closer inspection reveals a heritage bred of competition and luxury. The four huge disc brakes glinting behind wire wheels are probably the best visual definition of safety a state trooper will ever see. It doesn't take a trooper's expert eye though to sense the comfort and quality of the deep english leather bucket seats, woodrim racing wheel, and complete instrumentation in the sumptuous cockpit.

Assuming you don't have a badge, motorcycle and apologetic grin, we suggest you drop in at your local "Total Performance" Ford Dealer and spend a few concentrated moments carefully inspecting the world's most versatile sports car! $5995.00 P.O.E.

COBRA
POWERED BY FORD

FOR MORE INFORMATION WRITE: SHELBY AMERICAN, INC., 1042 PRINCETON DR., VENICE, CALIFORNIA

CORVETTE vs. COBRA:
the battle for supremacy

CONTROVERSY is the lifeblood of automobile racing, and the sport has recently been given another of its frequent transfusions. The big battle now being waged is between factions in the AC Cobra and Corvette Sting Ray camps, with the former's shouts having a decided ring of triumph and the latter's falling about midway between honest outrage and sour grapes. It seems that in a ridiculously short time, the Corvette has been clouted from its position of absolute primacy in large-displacement production-category racing, and Corvette fanciers are a trifle reluctant to accept the new state of affairs. Boosters of the Cobra (few of whom actually have any hope of becoming owners) are the people who have long been annoyed to see those big, ostentatious Corvettes drubbing the *pur sang* imported sports cars. The undeniable fact that the Cobra is as much *bar sinister* as *pur sang* does not appear to bother this group much; the Cobra looks every inch the traditional hand-built sports car (which it is, to a remarkable degree) and that is enough. In any case, the battle waxes furious, and emotional, and it is, therefore, interesting to examine some of the facts in the matter.

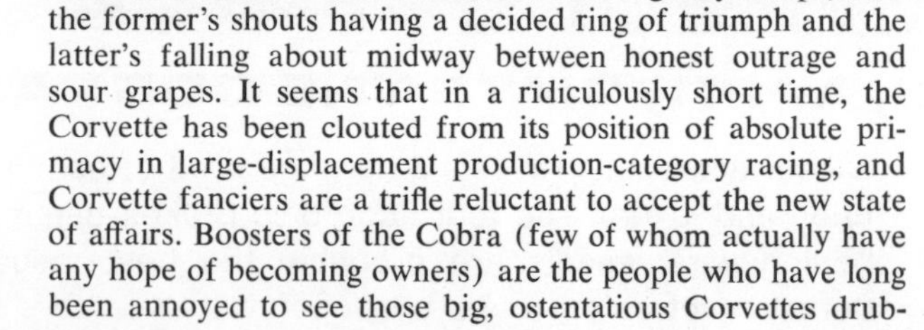

When comparing standard street versions of the Corvette and the Cobra, one can see the makings of rather an uneven contest. The Cobra has a curb weight of only 2020 lb, and the latest Ford engine used as standard in the car, the 289-cu-in. Fairlane V-8, has 271 bhp at an easy 6000 rpm. The Corvette presents a slightly confused picture, insofar as the touring version is concerned, because it is offered with engines in several states of tune. However, that most nearly comparable is the one having an engine equipped with the big, 4-throat carburetor, which gives it 300 bhp to propel its 3030 lb. Thus, the "average" Cobra one finds on the street will have a weight to power ratio of 7.45:1, while its Corvette counterpart, even though having more power, is heavier and has a less advantageous ratio of 10.1:1. Moreover, even if the Corvette purchaser is willing to go "whole-hog," and opt for the 360-bhp engine, he will still be hauling about 8.4 lb per bhp. The results are exactly what theoretical considerations predict. The "showroom-stock" Cobra will cut a standing-start ¼-mile in 13.8 sec, with a terminal speed of 113 mph, while a Corvette, in similar tune, is about a full second slower and will reach not quite 100 mph at the ¼-mile mark.

In top speed, too, the Cobra has the advantage. Its nominal frontal area of 16.6 sq ft gives it quite an edge on the Cor-

SHELBY AC COBRA: all-aluminum, hand-built coachwork on a chassis having a 90-in. wheelbase and powered by a Ford V-8 engine developing up to 340 bhp. It is less than 152-in. long, overall, and weighs only slightly over 2000 lb at curbside, ready to go.

vette, which is pushing away at 19.3 sq ft of air, and the touring version of the Cobra will exceed 150 mph (urk!), about 10 mph faster than the Corvette—even when the Corvette has the "big" engine. This disparity in top speed will continue, in all likelihood. The airflow over the Cobra is probably not as clean as that over the Sting Ray coupe, but the Cobra's advantage in frontal area cannot be denied. To counter that advantage, the Sting Ray would have to be 14% "cleaner" than the Cobra—and it isn't.

In handling, the two cars are more evenly matched than in any other area. Both cars have all-independent suspensions, and any advantage its lightness might give the Cobra in cornering power is just about offset by its rather primitive suspension layout—the Sting Ray has a much more sophisticated suspension.

The Cobra's basic chassis and suspension were laid down back in 1952, or thereabout, by Tojiero, in England, for a series of very limited production sports/racing cars. These were quite successful, and the design was bought by AC and adopted for its 1954 Ace sports/touring car. The Tojiero design, which borrowed heavily from Cooper's serendipitous Formula III car, has a frame that consists of a pair of large (3-in.) diameter steel tubes, with appropriate cross-bracing, and tall box structures at the chassis ends that carry the suspension elements. These elements are a transverse leaf spring, mounted atop the box structures, with a pair of A-arms underneath, giving an essentially parallelogram geometry and a roll center at ground level. This theme is repeated at both front and rear of the chassis.

With this suspension, the Cobra's wheels tilt with the chassis during cornering, and assume a camber angle that adversely affects cornering power. To compensate, the rear wheels, particularly on the competition Cobras, are given a fairly considerable amount of initial negative camber, so that the "outside" wheel is brought upright as the chassis leans, and that restores much of the tire adhesion that would otherwise be lost. Unfortunately, the tires are cambered too much for the best possible grip under straight-line acceleration. And this is no mere theoretical probability; the competition Cobra is notable for the difficulty it has in getting all of its thunderous horsepower applied to the road surface.

The Corvette Sting Ray, on the other hand, is a very recent design, and incorporates much of what has been proven desirable, in general suspension layout, over the past 3 or 4 years. It has the unequal-length A-arm (with coil springs) front suspension that has, with good reason, become standard for both passenger and racing cars, and a Lotus-inspired unequal-length link rear suspension. The roll centers are at a more modern height than is true of the Cobra, 3.25 in. in front and 7.56 in. at the rear. This, in itself, means that the Corvette will tend to lean a bit less than the Cobra, but the really important thing is that the outside wheels are held in a substantially upright attitude as the chassis leans, and the tires maintain good contact with the road. Also, the front suspension has its members angled upward to provide an anti-dive factor of about 50%, which, of course, cuts nose-dip under braking to half of what it would be without this feature. Finally, somewhat softer springs and longer wheel travel are provided in the Sting Ray's suspension, and the car rides more comfortably than the Cobra—which is, itself, not bad in that respect.

We would say that, in the touring versions, the Cobra and Corvette handle about equally well, with a slight nod in the Cobra's direction because of its lower bulk, weight and quicker steering. However, the Cobra's quick steering, now a rack-and-pinion setup in place of the former cam-and-roller steering box, is not entirely a blessing. The completely reversible nature of the steering box delivers road shocks from the tires right through, undiminished, to the steering wheel, and there are times when cornering hard when wheel-fight can be something of a bother. Here again, the Corvette also has its troubles: its steering, although accurate and free of feed-back, is just a shade too slow, and it is sometimes difficult to wind-on opposite lock fast enough to catch the car's tail as it swings out under a too enthusiastic application of power.

With regard to brakes, the Cobra scores heavily over the Corvette—at least insofar as sheer resistance to fade is concerned. Actually, disc brakes have not yet proven to be as trouble-free in day-in, day-out service as the better drum-type brakes, which the Corvette has.

Taken as touring cars, and bearing in mind all of the factors of reliability, service life, availability of service, comfort, utility, and that most important of intangibles, driving pleasure, it is difficult to make a choice. The Cobra is nominally an import, but the major mechanical elements are American manufactured, and most service problems can be handled by

CORVETTE STING RAY: futuristic fiberglass panels on a chassis having a 98-in. wheelbase and powered by a Chevrolet V-8 engine developing up to 360 bhp. It is a trifle more than 175-in. long, overall, and weighs slightly over 3000 lb at curbside, ready to go.

CORVETTE vs. COBRA:

any Ford garage. It is not outstandingly comfortable, if you happen to be talking in terms of driving from New York to Miami, and not about a sporting afternoon on mountainous back-country roads. Conversely, the Cobra is a somewhat more sporty machine on those same twisty roads than the Corvette. As has been said about so many places, the Cobra cockpit is a great place to visit for fun, but you wouldn't want to live there. As for trunk space, there isn't enough in either of the cars under discussion to argue about.

The Cobra's and Corvette's relative suitability as racing cars is seen in their competition records. Their first meeting, at Riverside Raceway last October, was inconclusive, as the Cobra was then only slightly faster than the "prodified" Corvettes running there, and the Cobra took a narrow lead only briefly, to retire immediately with a broken rear stub-axle. Shortly thereafter, the rivals met again, at Riverside once more, and on that occasion the domination of Corvette in its racing category came to an end. Dave MacDonald and Ken Miles, driving Cobras, beat all of the Corvettes (and there were some good ones there) so badly that it was not even a contest. Indeed, just to add insult to injury, Ken Miles made a pit stop after his first lap, ostensibly to have the brakes, or something, inspected, and after all of the Corvettes had gone by, he set out in pursuit. Whittling away at the Corvettes at the rate of about 5 sec per lap, on a 2.6-mile course, Miles caught his teammate, MacDonald, and relegated the first Corvette to 3rd-place in what seemed like no time at all.

The next confrontation was at the Daytona 3-hour, where a vast comedy of errors prevented the Cobras from defeating the GTO Ferraris (even though they demonstrated that they had the necessary speed) and Dick Thompson, in a Sting Ray, beat back the faltering Cobras to one-up them in that race. In the very recent Sebring Enduro, neither the Cobras nor the Corvettes fared particularly well. A rash of broken engines, and one transmission, eliminated 4 of the 7 Corvettes entered, and one of those still running at the race's end had been in the pits for a majority of the 12 hours having its engine bearings replaced. This Corvette completed only 46 laps.

The showing down at the snake (Cobra) pit was a little, but not much, more impressive; they lost exactly half of the 6 cars entered, and all of the finishing Cobras had to be nursed back from the ranks of the walking wounded at least once during the race. Even so, the Cobras' showing was better than the results indicate. Most of their problems were of a relatively minor nature (no shattered engines or other major components, at any rate), and while they were out on the course the Cobras showed more sheer speed than almost anything there. Phil Hill was observed, in practice, engaging one of the "prototype" Ferraris in a drag race up the pit straight and the good Phil, smiling hugely and rowing away at the gear-lever, carried it to a draw going into the first turn—after which the Ferrari moved away in no uncertain fashion. The Cobras, while they were in action, had plenty of speed, and the best-placed Corvette finished 10 laps behind the first of the Cobras.

One of the more interesting aspects of the great Cobra-Corvette debate is that the "Chevrolet-Forever" contingent has been complaining bitterly about the "unfair advantage" Shelby has taken in securing a list of approved competition options for his Cobras. This is indeed curious, for the ploy under attack is precisely the one used by GM to make its Corvettes competitive. In fact, we can draw parallels between almost every option offered for both cars. The Corvette has its fuel injection; the Cobra a double brace of 48-mm, double-throat downdraft Weber carburetors. Both have optional competition brakes with friction material largely unsuited to street-type driving. Aluminum alloy, cross-flow radiators are offered for both, as are competition exhaust systems, and cast light-alloy wheels can, due to a relaxing of the production car racing rules this year, be used on any car. Special, and very stiff, springs are catalogued for each car, as are dampers, and there are the miscellaneous items like oversized fuel tanks, for distance events, and more axle ratios than anyone could hope to need for either car. Transmission ratios? They are identical, each car using the same Warner Gear transmission. The Corvette is delivered with the close ratio gears for this gearcase installed as standard, and the wide ratio gears are offered as an option; the Cobra comes standard with wide ratio gears and the close ratio set is available as an option.

In full racing trim, both the Cobra and the Corvette would be thoroughly unpleasant to drive down to the office. The hot-cam, fuel-injected Corvette engine rumbles and chuffs smoke at low speeds, and so does the 340-bhp (at 6500 rpm), Weber-carbureted racing engine in the Cobra. Clutch and brake pedal pressures in both cars are fierce, and the low-end throttle response is awful. The major sin of the Cobra, in the Corvette booster's eyes, is that it is a winner, and it is likely to stay one unless a lightweight version of the Corvette is introduced. These cars' merits as touring machines can be argued, but there is no disputing which is the better racing car. The Cobra's lightness allows it to accelerate and corner faster, and stop quicker (primarily due to the advantage provided by its disc brakes), and on a straightaway of a likely length, the Cobra will be a good 10-mph faster. Given those points, it is very hard to imagine that any well prepared, well driven Cobra will be beaten this year—not by the Corvettes, and possibly not by anyone, unless the organizers get sneaky and push the Cobras over into the same races with all-out racing cars. There are, as a matter of fact, rumors of this happening, and if it does, the Cobras just might beat the big modified cars, too.

No matter where they run, the spectators will be the winners, for the Cobra is fast, noisy and slides about in a spectacular manner, and everyone will eventually learn to admire it for the tremendous sporting/racing machine it is—even the people who drive up to the spectator gate in a Corvette.

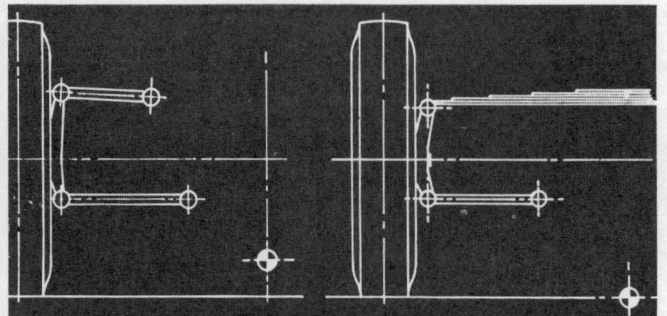

The Corvette front suspension, left, is more modern and has a higher roll-center than the Cobra's, at right.

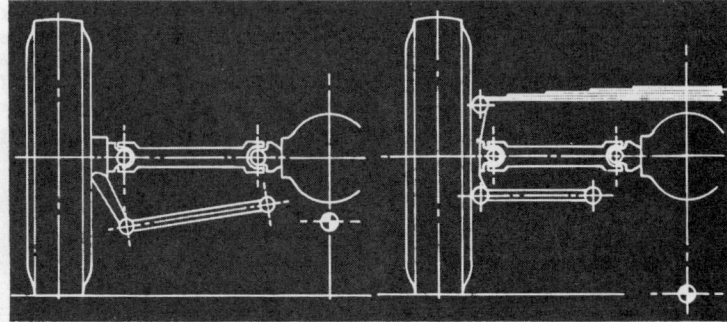

The Corvette's rear suspension has a higher roll-center than that of the Cobra, which as at the front, uses its spring as a link.

ROAD TEST
AC COBRA

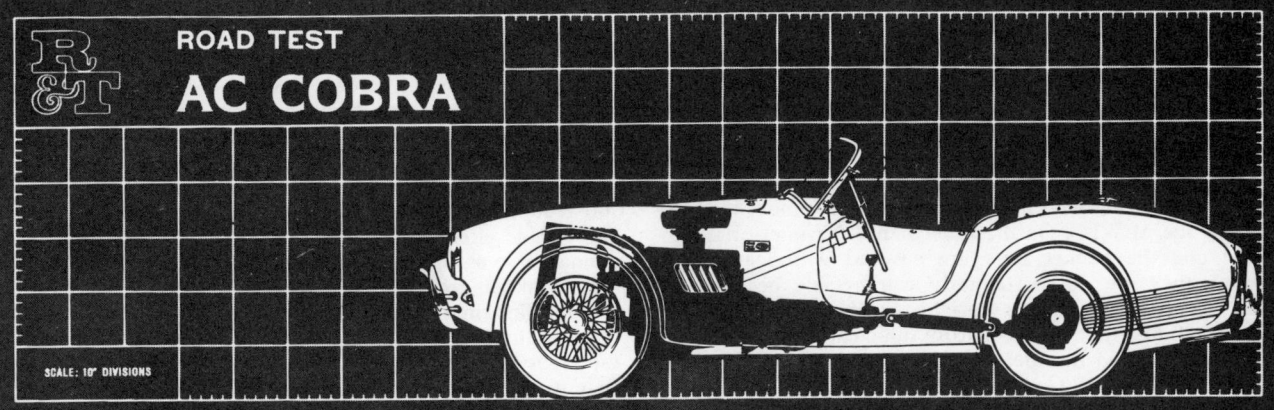

SCALE: 10" DIVISIONS

PRICE
List, West Coast POE.......$5995
As tested, West Coast.......$6343

ENGINE
Engine, no. cyl, type.....V-8, ohv
Bore x stroke, in.......4.00 x 2.87
Displacement, cc..........4730
   Equivalent cu in..........288.5
Compression ratio..........11.6:1
Bhp @ rpm..........271 @ 6000
   Equivalent mph............121
Torque @ rpm, lb-ft..314 @ 3400
   Equivalent mph.............69
Carburetor, no., make.1-4-bbl Ford
Type fuel required......Premium

DRIVE TRAIN
Clutch diameter & type:.....10.41
   Semi-centrifugal
Gear ratios, 4th (1.00)........3.77
   3rd (1.41)................5.32
   2nd (1.78)................6.71
   1st (2.36)................8.90
Synchromesh.............on all 4
Differential ratio............3.77

CHASSIS & SUSPENSION
Frame type: large diameter, ladder-type tube frame.
Brake type..................disc
   Swept area, sq in......est. 580
Tire size.............7.35 x 15
   Wheel revs/mi.............788
Steering type.......rack & pinion
   Turns, lock to lock..........2.0
   Turning circle, ft...........34
Front suspension: Independent with A-arm, transverse leaf spring, tube shocks.
Rear suspension: Independent with A-arm, transverse leaf spring, tube shocks.

ACCOMMODATION
Normal capacity, persons........2
Hip room, in............2 x 16.5
Head room.................35.5
Seat back adjustment, deg......0
Entrance height, in............41
Step-over height..............14
Floor height................10.5
Door width.................29.5
Driver comfort rating:
   for driver 69-in. tall........65
   for driver 72-in. tall........55
   for driver 75-in. tall........50

GENERAL
Curb weight, lb.........2170
Test weight.............2540
Weight distribution
   with driver, percent.....47/53
Wheelbase, in..............90.0
Track, front/rear......51.5/52.5
Overall length............151.5
Width.....................61.0
Height....................49.0
Frontal area, sq ft.........16.6
Ground clearance, in........5.0
Overhang, front..............30
   Rear......................39
Departure angle, no load, deg..18
Usable trunk space, cu ft.....5.5
Fuel tank capacity, gal........18

INSTRUMENTATION
Instruments: 160-mph speedometer, 8000-rpm tachometer, oil temp, water temp, oil pressure, fuel, clock, ammeter.
Warning lamps: turn indicators, generator, high beam.

MISCELLANEOUS
Body styles available: roadster as tested.

ACCESSORIES
Included in list price: 271-bhp engine, 4-speed gearbox, full instrumentation, wire wheels, limited-slip differential.
Available at extra cost: grille guard, wind wings, visors, heater, seat belts, luggage rack, chrome wheels, outside mirror, radio—plus many performance options.

CALCULATED DATA
Lb/hp (test wt)..............9.4
Cu ft/ton mi...............196.9
Mph/1000 rpm (4th).........20.2
Engine revs/mi.............2970
Piston travel, ft/mi.........1421
Rpm @ 2500 ft/min.........5230
   Equivalent mph.............103
R & T wear index...........42.2

MAINTENANCE
Crankcase capacity, qt..........5
   Change interval, mi........5000
Oil filter type.............paper
   Change interval, mi........5000
Lubrication grease points.......8
   Lube interval, mi..........1000
Tire pressures, front/rear, psi..28

ROAD TEST RESULTS

ACCELERATION
0–30 mph, sec.................2.2
0–40 mph......................3.4
0–50 mph......................5.0
0–60 mph......................6.6
0–70 mph......................8.6
0–80 mph.....................10.8
0–100 mph....................14.1
Passing test, 50–70 mph.......2.5
Standing 1/4 mi, sec.........14.0
   Speed at end, mph........99.5

TOP SPEEDS
High gear (6900), mph........139
3rd (7000)...................100
2nd (7000)....................79
1st (7000)....................60

GRADE CLIMBING
(Tapley Data)

4th gear, max gradient, %.....28
3rd.................off scale
2nd.................off scale
Total drag at 60 mph, lb.......115

SPEEDOMETER ERROR
30 mph indicated.....actual 31.0
40 mph.....................41.3
60 mph.....................62.6
80 mph.....................83.0
100 mph...................104.4

FUEL CONSUMPTION
Normal range, mpg........13–18
Cruising range, mi......230–320

ACCELERATION & COASTING

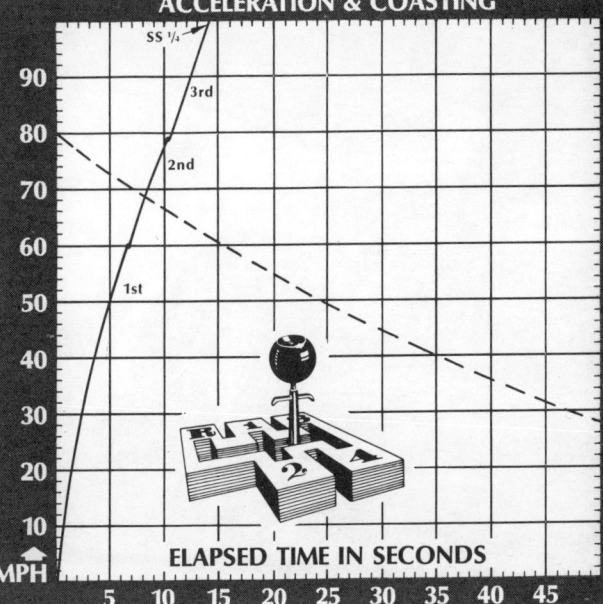

ELAPSED TIME IN SECONDS

15

and handling characteristics. However, the standard of finish of both the exterior and interior are very good and the aluminum body shows evidence of a considerable amount of hand workmanship. Unfortunately, the car does not offer very much in the way of creature comforts for its $6000. There are no windup windows and the top is a tent-like affair which one assembles around a collapsible frame carried in the trunk. The result is minimal weather equipment reminiscent of the MG-TC. On the other hand, the interior is carpeted and the bucket seats are covered with leather, although they are not particularly comfortable because the back seems to be far too upright. Furthermore, the taller driver will find that there is hardly sufficient room to operate clutch, brake and throttle.

On the road the Cobra is a curious mixture of ancient and modern. It is without doubt one of the fastest cars we have ever tested, and one can forgive almost anything for the sheer exhilaration of its performance. On the other hand, the suspension and steering reminded us of the sports cars we were driving 10 years ago, which is not necessarily a criticism. In an age when the sports car is expected to offer all the comforts of a family sedan, it is pleasant to revert to something like the Cobra which is nothing more than a weapon designed specifically for proceeding from one point to another in the minimum amount of time.

At low speeds both the suspension and steering are stiff, and it is not possible to tell that the rear suspension is independent. However, when the car is being driven in the manner for which it is designed, the road holding, steering and directional stability are very good. The behavior of the suspension is distinctly peculiar when the car is driven hard because the parallelogram layout at front and rear causes the wheels to lean excessively when cornering. This is particularly noticeable if one watches one of the competition cars during a race, however, the effect on the road holding is not as bad as it appears at first sight.

To compensate in part for the peculiarities of the suspension, the street Cobras are equipped with 7.35 x 15 Goodyear tires (Shelby handles Goodyear racing tires as a sideline), and apart from improving the road holding by the simple method of putting more rubber on the ground, they are a great asset to a car which needs all the traction it can get. This is carried a stage further in the competition models which are equipped with monstrous 8.20 x 15 Goodyear Stock Car Specials. Another noticeable feature of the street machines when viewed from the rear is the 3 degrees negative camber designed to counteract a tendency toward oversteering.

Although oversize tires are a definite improvement, they do tend to make the steering heavy at low speeds, but another side effect is that they contribute directly to the cars remarkable stopping ability which is insured by the Girling disc brake at front and rear.

Our first impression of the Cobra on the road was one of blinding acceleration in all gears. This was assisted considerably by the 3.77 axle ratio, and our feeling was that the car would be improved by a slightly lower numerical ratio because the engine is turning over unnecessarily fast at 70-75 mph cruising speeds. However, even with this ratio and using a peak of 7000 rpm, 60 mph can be achieved in 1st and a useful 100 mph in 3rd, and in consequence the passing ability of the car at any speed is nothing short of sensational. In traffic it is quite docile, although the clutch is heavy and considerable use must be made of the transmission.

Those people who are unaccustomed to driving cars with the potential of the Cobra would be advised to proceed with caution, because it is easy to find oneself going much too fast for comfort before one has realized what is happening. This does not mean that the car is particularly difficult to drive, because the tires and suspension do not permit excessive wheelspin off the line, unless one is very heavy handed, and shifting presents no problems at all. The main thing about it is that the Cobra reaches 100 mph while more conventional cars are struggling up to 75 mph, and this can be disconcerting to the uninitiated.

The AC Cobra is not just another sports car which someone has dumped a big V-8 into, but a properly conceived and developed machine which has been greatly improved by an extensive racing program. It offers exceptional performance without the problems of parts and service normally associated with cars of this nature. The AC Cobra is a sports car in the true sense of the term and, for those people who are not unduly concerned with comfort, it is a good value for the money.

Cobra treatment of the AC body gives the car a low-hung, rugged appearance.

Forty inches high, 350 horses strong, 200 mph fast—in other words, Total Performance!

FORD GT

Out of Dagenham and Detroit, on a Broadley-to-Wyer-to-Lunn triple play, a new racing car

WHEN THE FORD MOTOR Company returned to motor racing, it did so on a scale that is only possible for a vast corporation with an almost unlimited budget. So far, we have seen Ford powering Lotuses at Indianapolis, Ford at road races powering AC Cobras, and Ford at stock car races powering Fords. However, because it is an international company, Ford is now planning to contest international events, and the latest addition to its stable is the Ford GT.

The Ford GT looks suspiciously like last year's Lola and it is in fact a Lola derivative because most of the work was carried out at the Lola facilities in England under the direction of Eric Broadley. However, the car is called the Ford GT and the Lola name has been dropped. Others connected with the GT include Roy Lunn, the project manager, and John Wyer, previously a director of Aston Martin and now in charge of the racing program.

This is a strong team, particularly when one remembers that all the might of FoMoCo is behind it. However, it will need all the strength it can muster, because its intention is to oppose Ferrari in the prototype class of the Manufacturer's Championship. At the same time, the AC Cobras will be contesting the GT class against the Ferrari GTOs.

The GT was completed in a period of 12 months since its inception, and the whole project appears to have been carried out with an exceptionally high degree of planning and development work, utilizing all available resources such as wind tunnels and computers, and nothing seems to have been left to chance.

The whole conception of the GT is quite radical, the most striking feature of the car being its low height of 40 in. Other dimensions are: wheelbase 95 in., length 159 in. and track 54 in. The weight of the car less driver and fuel is 1820 lb and the distribution front and rear is 43-57%, which at first might seem to be an undue rearward bias. However, weight distribution of this nature is not uncommon in race cars, and can be compensated for by tires and suspension.

The main body section is a semi-monocoque structure and all the unstressed parts (such as the doors) are thin fiberglass fabrications. Much attention has been given the aerodynamics because the car is expected to approach 200 mph on the Mulsanne straight et Le Mans. Not only did this involve extensive wind tunnel testing of the body shape itself, but also the air flow to the various components requiring cooling,

CONTINUED ON PAGE 85

The three parents—John Wyer, Eric Broadley and Roy Lunn.

Weber carbs and bundle-of-snakes exhaust on '63 Indy engine.

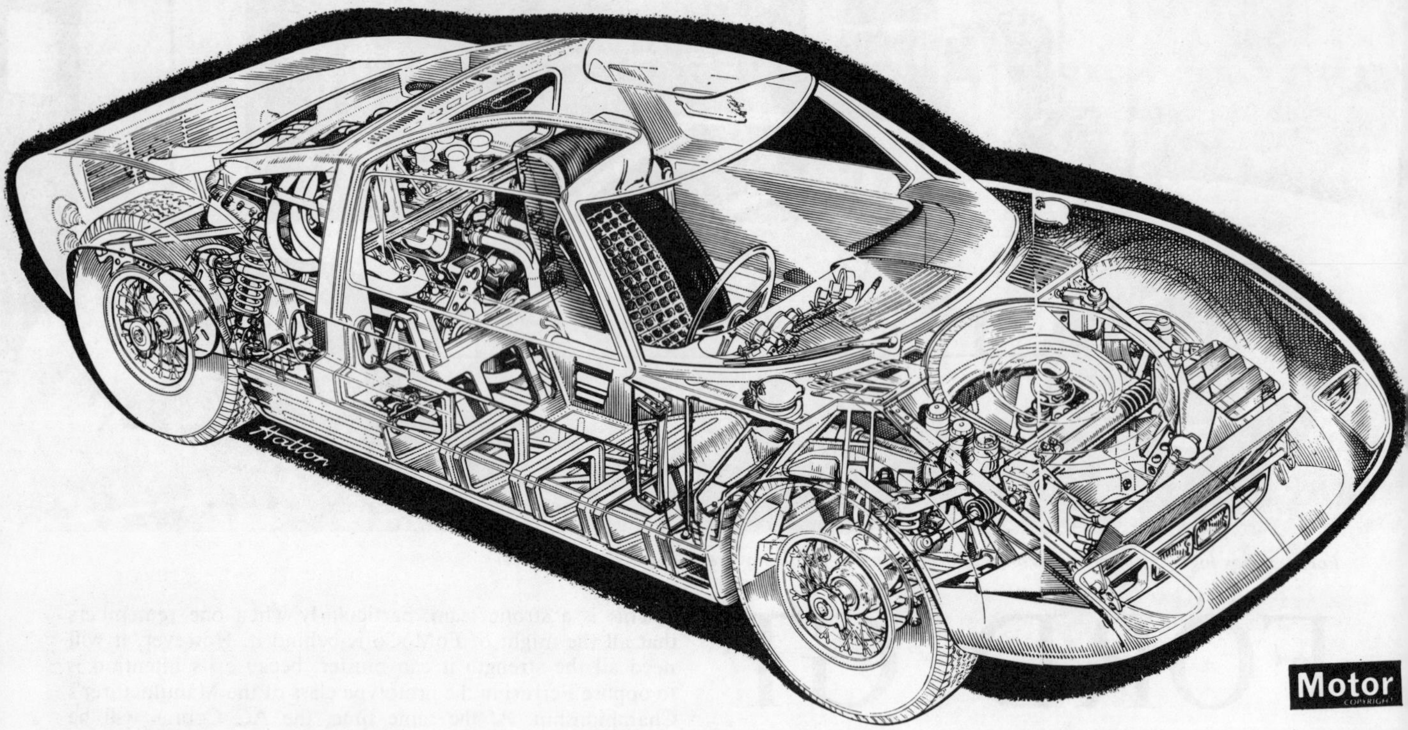

FORD GT

Ford's exciting prototype comes a cropper in the rain at the Le Mans trials

THE FORD GT, details of which were printed in the June issue, made its first race circuit appearance at the Le Mans trials on a rainy Saturday in April. And a pretty sad Saturday it was, too, for the "Total Performance" image of the Ford Motor Co.

Almost before any idea could be gotten as to how the car was going to perform, the two cars that were taken to the site of the famed 24-hr race had been pranged sufficiently that further laps were out of the question. First, French driver Jo Schlesser lost one in a fast bend on the long straight,

CONTINUED ON PAGE 35

"My inflating seat's gone stark, staring bonkers!"

G.T. 350

Precise control takes on a new meaning behind the wheel of Shelby American's new Mustang GT-350! The complete suspension system has been re-designed to comply with a computer plotted geometry that allows the GT-350 to stick like nothing you've ever driven. Reflex quick steering, Koni shocks, and Goodyear 130 mph "Blue Dots" let you pick your exact line... and stick to it! No use mentioning the GT-350's 306 hp Cobra hi-riser 289 or the competition proven disc brakes until you've actually sampled a few corners at speed. The most complex blind apex closing radius bend becomes the expert driver's challenge instead of an exercise in frustration. When you go down to test the GT-350, ask the salesman to bring your present car along for comparison... you won't believe it! Suggested list price $4547.00

for more information write Dept S, Shelby American, 6501 W. Imperial Hwy., Los Angeles, California 90009

ROAD TESTS: MUSTANG 350-GT, NEW TRIUMPH TR-4A

ROAD & TRACK

MAY 1965 THE MOTOR ENTHUSIASTS' MAGAZINE 3/6 IN ENGLAND 60¢ IN CANADA 50 CENTS

CARROLL SHELBY & THE CARS HE WILL RACE IN 1965
427 COBRA, MUSTANG 350-GT & THE FORD GT COUPE

The Colorful, Exciting & Slightly Mad World of Rallying

ALICE BIXLER PHOTO

A Look at the Daytona Winner FORD GT-40

BY TONY HOGG

UNFORTUNATELY, the inner sanctums of Detroit, where top level decisions are made, are barred to members of the fourth estate, and even to the staff of *Road & Track*. It is therefore impossible to tell when, why, and by whom the decision was made at Ford to project an image of "Total Performance," and engage in a comprehensive and systematic program of racing to include such events as Le Mans, Sebring and similar international races in the prototype class.

At one time there was some speculation that Ford would buy out Ferrari lock, stock and barrel, and nasty-minded people felt that this was based on the theory that "if you can't beat 'em, buy 'em." Apart from any difficulties that undoubtedly arose during the negotiations, it is much more likely that Ford's decision was based on the fact that the resulting cars would be nothing much more than Ferraris bearing the Ford nameplate, which would have done little or nothing for the Ford image.

One of the reasons for the negotiations with Ferrari was to accelerate the program by taking over an existing enterprise, rather than starting completely from scratch. But when the deal fell through Ford began looking about for someone to build cars for its project, finally picking Englishman Eric Broadley, whose Lola GT coupe was very similar to designs on the boards at Ford, but was already in the testing stages.

Broadley is a construction engineer by training, who got into race car building as an avocation. Quiet and unassuming by nature, he is probably quite the equal of Colin Chapman as an engineer and, although he has received little credit for the project, he is the father of the Ford GT.

The next step was to undertake an extensive program of testing and research on the existing Lolas. This was conducted at various English circuits and at Monza, using drivers such as Roy Salvadori and Bruce McLaren. Meanwhile, John Wyer, who had managed the Aston Martin racing team for a number of years, was put in charge of development in England, and Englishman Roy Lunn, who has been with Ford of Dearborn for several years, was appointed chief designer.

The new car was officially presented to the press for the first time at the 1964 New York Automobile Show, where it was introduced by Ford vice president Lee Iacocca. During the course of his speech, Iacocca referred to it as "an American Manufacturer's car," which is one way of describing it, although, apart from the engine and one or two very minor items, it would appear to have been built in England by Englishmen and designed in America by Americans led by an Englishman.

The 1964 season was not particularly successful and the car did not win any races. However, it created great interest and a very favorable impression wherever it appeared. A lot was learned by the team, and Richie Ginther was timed officially at 207 mph on the Mulsanne straight at Le Mans. During 1964, Carroll Shelby was also campaigning his Cobras at the same events so, in fact, Ford was competing against itself with two totally different teams.

In order to unify the road racing project and bring everything under one roof, a decision was made to hand the GT project over to Shelby American for 1965, but to continue construction of the cars at Ford Advanced Vehicles, at Slough, England. Phil Remington, Carroll Shelby's chief

In the latest version of the Ford GT, a Cobra-modified Ford 289-cu-in. engine is used. The engine oil and differential oil coolers are placed on either side of the Type 37 Colotti transmission and fed by ducted air.

FORD GT-40

engineer, had been closely associated with the project since its inception so he has been able to continue development work without the necessity of commuting to Dearborn or London.

At this time, some 15 GTs have been built, and present plans call for a total of 50, some of which will be open cars, some coupes, and some will be set up for street use. Work is also under way on a 427-cu-in. prototype. For those people who are in the market for this type of fast transportation, the price has not yet been fixed, but it is rumored to be in the region of $15,000.

When the cars were received by Shelby, they were put through a series of exhaustive tests at Riverside and Willow Springs. From these tests it was apparent that several problems existed, and the major ones related to the engine temperature, the transmission, the brakes, the weight, and the road-holding at very high speeds.

Fortunately for the Shelby organization, the Aeronutronic division of FoMoCo (in Newport Beach, Calif.) numbers among its employees several racing enthusiasts who are also experts in the instrumentation and aerodynamics of missiles. By enlisting their aid and equipment, a lot of progress was made on these problems in a comparatively short time.

The engineers from Aeronutronics were amazed at how little was known about what they referred to as "low speed aerodynamics," as applied to 200-mph race cars. This term would seem reasonable enough for people engaged in the 18,000-mph nose cone business. In fact, aerospace techniques have proved very successful when applied to race cars and it is likely we will see considerably more progress in this particular area in the future.

When the Ford GT was in the development stage, a good deal of reliance was placed on wind tunnel testing for air flow over the body and also through the various ducts. Although a large amount of data was obtained from these tests, the car didn't respond as the tests indicated it should.

It would appear that there are several reasons for this state of affairs, and the main one is that background data concerning the wind tunnel testing of cars are very limited. Second, little is known about reproducing ground effect in a wind tunnel, because it is not a factor in aircraft design, and third, scaling down a car to model size is relatively much more critical than with aircraft if the resulting car is to behave on the road as the model does in the wind tunnel.

With the assistance of Herbert L. Karsch, Loyle E. Baltz and Bob L. Pons, among others from Aeronutronics who donated their time, two different systems of testing were used, although both are basically the same. In the first, telemetry was employed so that information from instruments in the car was transmitted to a truck parked by the circuit, where data could be either read instantaneously from instruments in the truck or, alternatively, checked later from tape.

The second method was to mount an oscillograph recorder in the car to record data on paper with the car in motion. In each case, such information as the air pressures and temperatures in the ducts, the engine revolutions and, by coupling potentiometers to the suspension, the exact movement of the suspension at any point on the course could be recorded with the car being driven at racing speeds.

The advantages of using aerospace techniques are that time is saved, there is no hit-and-miss involved, and an

Rear suspension has outboard brakes; Metalastik joints inboard.

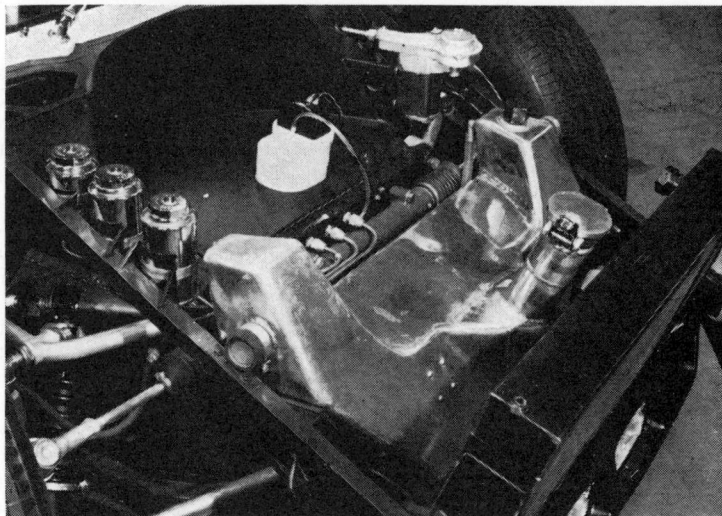

In the car's original form, four gallons of oil were carried in a tank mounted in the nose for the dry sump system.

Cockpit was engineered for maximum driver comfort. Sound level is low and interior remains cool.

exact record of the car's behavior under racing conditions can be compiled. On the other hand, the equipment is expensive and it requires skilled operators. However, it is ideally suited for the purpose because it is exceptionally light and compact, it will withstand all the shocks and vibration of a race car, and it can be run off the car's power supply because the oscillograph draws only about 8-10 amps.

Apart from the very sophisticated methods of testing introduced by Aeronutronics, much time was spent driving the car with tufts of wool attached to the body, and then following its progress around the circuit in a chase car carrying a passenger with a Polaroid camera.

With the experience gained during the 1964 season, and as a result of tests conducted by Shelby American, a number of changes have been made to the Ford GT. To overcome the water temperature problem, the nose has been redesigned and the ducting altered. In its original form, the engine was dry-sumped and four gallons of oil were carried in a front-mounted tank. By reverting to a wet sump, and eliminating the tank and its associated plumbing, about 75 lb were saved. At the same time, this left more room in the nose for a redesigned cooling system.

As far as the wet sump is concerned, it has to be extremely shallow and the capacity is nine quarts, but no oil temperature problems have occurred. An additional advantage of removing the oil tank from the nose was to permit the designer more latitude, because the original nose was found to be a near perfect air foil which caused the front of the car to lift at very high speeds. This has been corrected and the high speed stability has improved considerably.

The car was originally powered by the 4.2-liter pushrod Indy engine, but this has been replaced by the competition Cobra version of the Ford Fairlane 289-cu-in. unit. Although heavier, the 289 is cheaper and its 13% greater displacement provides more torque at a lower speed, which is an advantage because there are only four forward speeds in the Type 37 Colotti transmission. At the same time, the Shelby organization has had a vast amount of experience with the 289, so that the mechanics can set up an engine for a particular circuit and race distance.

During the 1964 season, trouble was experienced with the Colotti transmission-cum-differential. New, Ford-made ring and pinion gears gave a partial cure, and a more recent change is the adoption of involute splines for the four dog clutches which are used to effect gear changes. These changes are expected to cure most of their troubles, pending the arrival of all-new units from ZF in Germany. These will have five forward speeds, by the way. There is an oil cooler on each side of the transmission, with air supplied by special ducts. One cooler is for engine oil, the other for the transaxle lubricant.

The clutch is an English Borg & Beck 7.25-in. unit containing three driven plates. Its capacity is sufficient to handle the power output of the 289-cu-in. engine, although it would become marginal if the torque were increased by an appreciable extent. In order to reduce shock on the transmission, the inboard universal joints are Metalastik. These are a potentially weak point in the car (rubber) and they have been redesigned for longer life.

On circuits which place a premium on brakes, the Girling system on the GT has come very close to its limit on several occasions. The car was originally equipped with wire wheels in order to assist in brake cooling, but these did not help so a switch has been made to magnesium wheels, which are both lighter and more rigid.

At present, solid brake discs of 11.563-in. diameter are

Front suspension is extremely rugged.

FORD GT-40

used front and rear, with one piston on each side of the caliper operating one pad on each side. To improve cooling, the ducts to the brakes have been modified, and the next step is to switch to Kelsey-Hayes ventilated discs, which not only present a bigger surface area for heat dissipation, but also tend to draw air into the center of the disc and expel it at the periphery. Another modification to the braking system involves the use of Teflon lines covered with steel braid to prevent expansion due to the very high hydraulic pressure.

The rear brakes are mounted outboard, and whether this is from choice is not known but apparently there is insufficient clearance for inboard mounting. Aggravating the brake cooling problem is the size of the rear tires. Carroll Shelby has always favored getting plenty of rubber on the ground, so 9.00-15s are used at the rear mounted on 9.5-in. rims and the maximum section of these tires is approximately 12 in.

Since Shelby American took over the cars, the suspension has remained basically unchanged. Its chief characteristics are that it incorporates a considerable amount of anti-dive at the front and anti-squat at the rear, and it also appears to be abnormally heavy so that unsprung weight is greater than necessary. In fact, the whole car is probably about 200 lb above design weight, although wet-sumping the engine and eliminating the wire wheels have cut this to some extent.

From conversations with the drivers and after riding as a passenger in the car, we observed that the GT gets an awful lot of power to the ground as soon as the power is needed. In a slow turn, it tends to understeer slightly and, when entering a fast turn on a trailing throttle, it gets a bit twitchy. However, these are merely characteristics rather than faults. The absence of dive and squat are distinctly noticeable under conditions of heavy braking and acceleration.

In direct contrast to some other competition cars, driver comfort was obviously a concern of the designer, so that the GT can be driven for long periods without making undue demands on the stamina of the driver. The noise level is comparatively low inside the car, and the flow of fresh cool air through the driving compartment is carefully controlled. These two factors alone can make a big difference in a long race such as Le Mans (for which the car was designed).

In its original form, the car was equipped with inflatable seats which could be adjusted to suit the requirements of any driver. These have now been discarded on the grounds of undue complication rather than from fear of a blow-out, and the present seats are not adjustable but the pedals can be moved instead.

It is apparent that the Ford GT is one of the most sophisticated competition cars ever built. Admittedly, Parkinson's Law has been applied to some extent, as one would expect in such a big corporation, with everyone from the styling department to the janitor wanting to get in on the act. However, the basic design shows great potential and one cannot help feeling that Ford made a wise decision in electing to run during 1965 under the Shelby American banner.

MUSTANG GT-350

Shelby-American, transforms Ford's gentle little colt into a roaring, snorting stallion

WE SOMETIMES FIND it difficult to know when to take Carroll Shelby seriously. He's a great kidder. He has been known to put more effort into staging a really elaborate stunt than most people will to get rich. Therefore we're never really sure whether what he does is for real or is simply a result of his far-out whimsy.

The latest item we're not really sure about is the Ford Shelby American Mustang GT-350. The admitted purpose of the car is to win class BP in the Sports Car Club of America's production category racing. Which is a pretty amusing reason for building a car in the first place. Except that this isn't altogether a new car. It's a Ford Mustang with the 2+2 fastback body plus those alterations that Shelby American deemed necessary to outrace such cars as the pre-Sting Ray 283-cu-in. Corvette, the 3.8 or 4.2 XK-E Jaguar, the 260-cu-in. Ford-powered Sunbeam Tiger and miscellaneous others in the somewhat unreal world of SCCA production category racing. So why didn't Shelby American just fix up a few Mustangs for racing and go at it? First, to be assured of winning, the racing Mustang would resemble the Ford factory Mustang so little that it wouldn't be eligible for SCCA production racing. Which is all part of the joke, really, because SCCA insists that at least a hundred examples be "series-produced with normal road touring equipment" in a year. So Shelby undoubtedly let off a big whooping laugh and made plans to do exactly that—build a "street" version with normal road touring equipment and then tailor a competition version around that.

A brief run-down of the changes made in the standard Mustang will not only illustrate the thoroughness with which Shelby American went at the project but also show what was necessary to get the job done. To start with, the street version GT-350 has the same unit construction chassis and the same basic sheet metal as the 2+2 Mustang except for a fiberglass hood. It also uses the standard interior trim, front seats and instruments. Virtually everything else is either changed or completely different—sheet metal headers, increased oil capacity, wider wheels, high-speed tires, re-engineered front suspension, beefed-up rear suspension, limited-slip differential, Warner T-10 gearbox, no rear seat, spare tire moved to platform behind the seats, battery relocated into trunk, wood-rimmed steering wheel, added tachometer and oil pressure gauge, disc brakes at the front, heavy-duty drum brakes at the rear, quicker steering, Koni shocks all around and the whole package topped off with a special paint job. The competition version is basically the same, only more so—lightweight seat shells, stripped instrument panel, bare floors, rollbar, plastic windows, fiberglass front end section with built-in air scoops, wider wheels, different tires, bigger gas tank, bigger water radiator and an oil cooler for the differential. Get the picture?

The engine of the GT-350 is basically the 289-cu-in. high performance Ford engine but uses Ford's new high-riser manifold (which gives a tuned intake effect) with the new center-pivot float 4-barrel Holly carburetor which will not flood or starve during hard cornering. The GT-350 also uses lightweight tubular headers and straight-through mufflers. The engine is dressed up by the use of a thin air cleaner, handsome finned aluminum rocker covers and oil pan. The oil pan increases the sump capacity to 6.5 quarts

MUSTANG GT-350
AT A GLANCE...

Price as tested	$4584
Engine	V-8, ohv, 4736cc, 306 bhp
Curb weight, lb	2790
Top speed, mph	124
Acceleration, 0-60 mph, sec	6.8
Passing test, 50-70 mph, sec	3.4
Average fuel consumption, mpg	14

MUSTANG GT-350

(from 5.0) and includes baffles to assure that the oil doesn't surge away from the pickup. The engine is rated at 306 bhp at 6000 rpm, exactly 35 more than the 271 bhp figure advertised for the standard high-performance version.

Because weight is saved wherever possible in the GT-350, the Warner T-10 gearbox with aluminum case is used instead of the heavier Ford 4-speed unit. The limited-slip differential, made by Detroit Automotive Products and called the "No-Spin," is a heavy-duty unit which is used in trucks in civilian life.

The GT-350 uses the same basic suspension system as the standard Mustang but there are important differences. At the front, the inner pivot of the upper control arm has been moved down one inch. This results in greater changes in wheel camber during cornering, which keeps the front wheels more nearly vertical. It also raises the front roll center and consequently reduces the tendency to plow. Adding to the front roll stiffness, the diameter of the anti-roll bar has been increased from 0.84 to 1.00 in. At the rear end, the live axle is retained by 4-leaf semi-elliptics plus beefy torque reaction arms that sit atop the axle and are anchored into the chassis. Adjustable Konis are used at both front and rear and stiffer shock settings are reportedly the only difference in suspension between the street and comptition versions.

In appearance, the GT-350 is readily distinguished from the standard Mustang. First, all GT-350s are white with blue racing stripes. There is also the fiberglass hood with pin-lock hold-downs and the giveaway airscoop to clear the high-riser manifold. The air intake at the front is simplified by the use of an anodized grille and smaller horse than the decorative cross bars and insignia of the standard Mustang. And in case you overlook everything else, including the bigger wheels and the 130-mph rated 775-15 Goodyear "Blue Dot" tires, there's a "GT 350" painted on the lower panel ahead of the doors on either side of the car. The overall effect is good, we think, simple and uncluttered—and with room for big racing numbers.

The list price of the GT-350 is $4311 (plus $273 for the cast magnesium wheels if you want it to look like our test car) and it will be marketed through dealers who also handle Cobras. Shelby American is tooling up to produce as many

Comparing the street and competition versions

Street version of the GT-350 uses standard Mustang seats, soft trim, instruments, and adds wood-rimmed steering wheel plus tachometer and oil pressure gauge mounted on cowl.

Cockpit of the competition version has lightweight shell seats with deep sides, carpeting removed from floor, stripped dashboard and complete set of instruments.

as 200 copies a month for this trade. The competition version with the rest of the goodies will be sold only through the Venice, Calif. factory and will go for about $6000. These prices seem eminently reasonable considering the highly specialized and specifically tailored product that is being offered.

The driving position of the GT-350 is slightly better than in the standard Mustang, thanks to the nearly flat wood-trimmed Cobra steering wheel which has replaced the deep-dish Ford type. The long-legged driver still hits the turn signal lever with his knee when he puts his foot on the clutch, but except for this the driving position is good. The standard Mustang seats offer little lateral support but the extra-wide air force type seat belts clamp the hips firmly in place.

The standard key start is retained and once warmed up the engine settles down to a steady rumble at about 800 rpm. A touch on the throttle pedal results in a little twitch from the gyro effect of the engine and an impressive *harrumpha* from the exhaust pipes which exit ahead of the rear wheels. Perhaps it wasn't typical but the carburetion on our test car seemed a bit rich at the low end, tending to load up and make it necessary to clear its throat periodically when easing along in slow traffic. This attracted widespread attention, of course, and the teenagers who sidled up to eye the car gave us a hint how the fastest gun in town must have felt when he walked down main street in the old West.

Driving the GT-350 is as enlightening as it is dramatic. The clutch pedal is heavy compared with that of the standard Mustang but the action has a distinct "feel" and can be accurately controlled. This, plus the tight rear end that tends to stay on the ground rather than hop up and down, makes it an easy car in which to get good clean acceleration. You simply feed on enough throttle to break the tires loose, feather it slightly to pick up traction, then mash on it and watch the tach needle wind around toward the 6500-rpm red line. The gearbox linkage is excellent, the synchro faultless and you can chirp the tires at each shift if you want to. Our acceleration figures, obtained over the surveyed quarter-mile at Carlsbad Raceway, represent typical figures obtained in six timed runs. Our first try resulted in a flat 15.0 sec, very respectable indeed, and the two best runs were timed at 14.6.

The ride of the GT-350 would have to rate as poor if comfort were the main consideration. The springs, though

Street model GT-350 retains normal trunk trim and function except that spare has been removed and battery inserted.

Competition version has optional 37-gal. fuel tank in trunk. Big filler cap has funnel-type splash catcher.

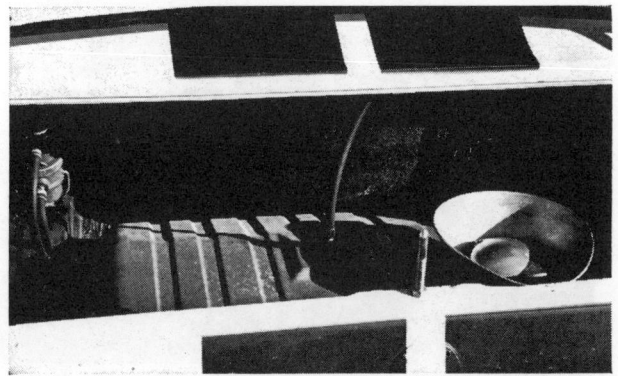

Ford 289-cu.-in. engine with new center-pivot float Holly 4-barrel is rated at 306 bhp at 6000 rpm in GT-350 form.

Engine in competition version uses funnel air intake from scoop in hood, stock rocker covers, bigger radiator.

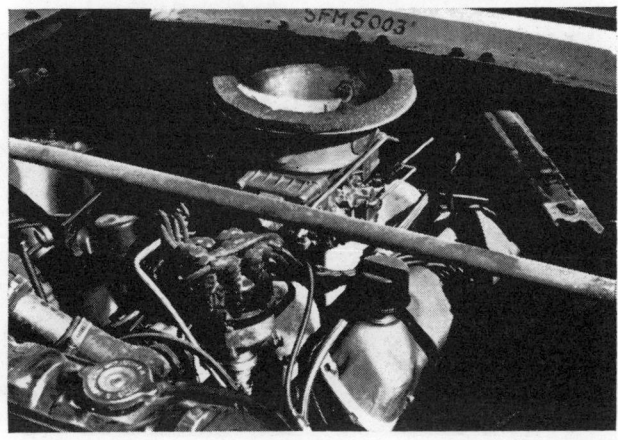

GT-350 uses 775-15 Goodyear Blue Dots on 6-in. wide rims.

Oil pressure gauge and 8000-rpm tachometer supplement standard instruments.

Covered spare tire is located behind seats on molded fiberglass platform.

MUSTANG GT-350

comparatively soft, are snubbed by the stiff shock settings and an abrupt dip results in thunking at both ends. On glass-smooth turns, the GT-350 is very fast, the considerable body lean not at all disconcerting to the driver and a cornering attitude can be readily maintained with the throttle. The most effective technique for a slow turn seems to be to wait late, brake hard, shift down, point the nose toward the apex and push it around with the throttle. This is easily controlled in the GT-350, even when the rear end begins to go and it's time to feather back a bit to keep the car aimed in the right direction. Past the apex, winding off, the GT-350 will take a surprising amount of throttle without losing its balance.

Over rough paving, the cornering technique is altogether different as the combination of power, mass and stiff suspension demands a very delicate touch. The rear end, which carries only 45% of the car's weight, even after all the changes which were made, is inclined to lose its poise on rough surfaces and changes in throttle opening must be made with extreme care if embarrassment is to be avoided.

The steering of the GT-350 has been quickened by lengthening the Pitman arm so the overall ratio is now 19:1 compared to the standard 27:1 and the turns lock to lock reduced from 5.0 to 3.75. You give a little for everything you get, naturally, and this quicker steering also results in more muscle being required to move the wheels.

Comparing

Plastic rear window is used in competition version, which not only saves weight but has space at top for ventilation.

Plastic windows in competition version permit weight saving as standard window winding mechanism can be removed.

Competition GT-350 has air ducts to brakes in lower nose section and also uses Goodyear T-7 racing tires on 7-in. rims.

Seats in competition GT-350 are thin-padded lightweight fiberglass shells which afford excellent lateral support.

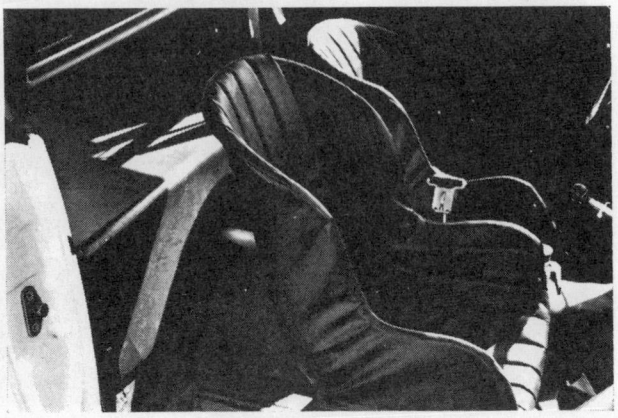

Rear brakes, 10-in. diameter x 2.5-in. width, use segmented metallic linings.

Torque reaction arm and 4-leaf semi-elliptics assure positive location of beam rear axle.

Front brakes are 11.375-in. Kelsey Hayes ventilated discs.

The brakes, which are 11.375-in. Kelsey Hayes ventilated discs in front and 10 x 2.5-in. heavy-duty metallic lined drums in back, are eminently suited to their job. The pedal pressure is a bit high, as you'd expect, but is in keeping with everything else about the car. The competition version uses the same brakes, of course, except that harder pucks are used for increased longevity.

The curb weight of our test car came out to 2790 lb, which is about 150 lb less than a standard Mustang, due primarily to the fiberglass hood, the lighter weight of the tubular headers and the omission of the rear seat. Even with the re-located battery and the weight lost toward the front, the weight distribution of the GT-350 is 55/45 front/rear compared with the 56/44 of the standard version. The competition version of the GT-350, which has an "approved" SCCA racing weight of 2550 lb (minus another 5% if it's needed), should get down toward the minimum with its more stark interior, plastic windows and so on.

All in all, the GT-350 is pretty much a brute of a car. There's nothing subtle about it at all. Making the obvious comparison to the Shelby American Cobra, or even the 325-bhp Sting Ray, the GT-350 seems more suited to the drop-out than the serious scholar. It will undoubtedly assure its owner of much attention whenever it is driven down the street, *rumphed* at a stoplight or parked at the drive-in. For the racing driver, it will also be a source of great amusement, as it should enable him to laugh all the way to the winner's circle in SCCA's class BP racing.

Exhaust vents behind doors are retained in street version but are removed and sealed over in competition car.

Cast aluminum oil sump increases capacity to 6.5 from 5.0 quarts.

Fiberglass molding around rollbar is typical of good workmanship.

Fiberglass hood with integral airscoop saves worthwhile amount of weight.

ROAD TEST
MUSTANG 350 GT

SCALE: 10" DIVISIONS

PRICE
List price..................$4311
Price as tested...........$4584

ENGINE
No. cylinders & type.....V-8, ohv
Bore x stroke, in......4.00 x 2.87
Displacement, cc..........4736
  Equivalent cu in............289
Compression ratio.........11.5:1
Bhp @ rpm..........306 @ 6000
  Equivalent mph............122
Torque @ rpm, lb-ft (est)
  ....329 @ 4200
  Equivalent mph.............86
Carburetors, no. & make.1 x 4 Holly
  No. barrels & dia.....4–1.562
Type fuel required.......premium

DRIVE TRAIN
Clutch type......semi-centrifugal
  Diameter, in..............10.4
Gear ratios: 4th (1.00)....3.89:1
  3rd (1.20)..............4.67:1
  2nd (1.62)..............6.30:1
  1st (2.36)..............9.18:1
Synchromesh............on all 4
Differential type......limited slip
  Ratio...................3.89:1
  Optional ratio..........4.11:1

CHASSIS & SUSPENSION
Frame type: welded platform with boxed side rails.
Brake type...........disc/drum
  Swept area, sq in........381
Tire size..................775-15
  Make & model.Goodyear Blue Dot
Steering type....recirculating ball
  Overall ratio.............19:1
  Turns, lock to lock......3.75
  Turning circle, ft........38.0
Front suspension: independent with upper A-arms, lower single arms, coil springs, tube shocks, anti-roll bar.
Rear suspension: live axle, semi-elliptic leaf springs, tube shocks, torque control arms.

ACCOMMODATION
Normal capacity, persons........2
Seat width, front, in........2 x 22
Head room..................39.5
Seat back adjustment, deg......0
Entrance height, in..........48.5
Step-over height..........14.25
Door width.................45.0
Driver comfort rating:
  For driver 69-in. tall........95
  For driver 72-in. tall........85
  For driver 75-in. tall........70
  (85–100, good; 70–85, fair; under 70, poor)

GENERAL
Curb weight, lb.............2790
Test weight.................3140
Weight distribution (with driver), front/rear, %....55/45
Wheelbase, in..............108.0
Track, front/rear........56.5/57.0
Overall length, in..........181.6
  Width....................68.2
  Height...................51.2
Frontal area, sq ft..........19.3
Ground clearance, in........5.5
Overhang, front/rear....35.0/41.5
Departure angle (no load), deg..18
Usable trunk space, cu ft.....8.5
Fuel tank capacity, gal......16.0

INSTRUMENTATION
Instruments: 8000-rpm tachometer, 120-mph speedometer, fuel, water temperature, oil pressure.
Warning lights: ignition, oil pressure, high beam, turn signals.

MISCELLANEOUS
Body styles available: fastback as tested.

ACCESSORIES
Included in list price: tachometer, oil pressure gauge, seat belts, fiberglass hood, limited slip, steel wheels, wood rimmed steering wheel, spare tire cover.
Available at extra cost: cast magnesium wheels, full range of competition options.

CALCULATED DATA
Lb/hp (test wt)..............11.0
Cu ft/ton mi.................157
Mph/1000 rpm (high gear)...20.3
Engine revs/mi..............2950
Piston travel, ft/mi.........1410
Rpm @ 2500 ft/min........5220
  Equivalent mph............106
R&T wear index............41.6

MAINTENANCE
Crankcase capacity, qt......6.5
  Change interval, mi......6000
Oil filter type............full-flow
  Change interval, mi......6000
Chassis lube interval, mi...36,000
Tire pressure, psi..........28-32

ROAD TEST RESULTS

ACCELERATION
0–30 mph, sec..............2.4
0–40 mph...................3.6
0–50 mph...................5.2
0–60 mph...................6.8
0–70 mph...................8.7
0–80 mph..................11.2
0–100 mph.................19.0
Passing test, 50–70 mph....3.4
Standing ¼-mi, sec........14.7
  Speed at end, mph........90

TOP SPEEDS
High gear (6100), mph.....124
3rd (6500).................110
2nd (6500)..................82
1st (6500)..................56

GRADE CLIMBING
(Tapley data)
4th gear, max gradient, %.....20
3rd.........................25
2nd...................off scale
1st....................off scale
Total drag at 60 mph, lb.....133

SPEEDOMETER ERROR
30 mph indicated....actual 26.4
40 mph....................34.5
60 mph....................50.4
80 mph....................67.0
100 mph...................83.2

FUEL CONSUMPTION
Normal driving, mpg......12–16
Cruising range, mi......190–250

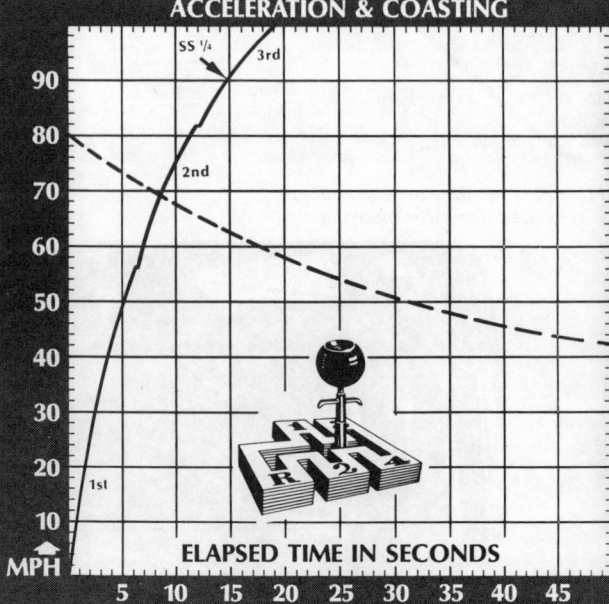

ACCELERATION & COASTING
ELAPSED TIME IN SECONDS

SHELBY G.T. 350 IS "SON OF COBRA"

ADULTS ONLY!

Shelby-American presents America's fastest, race-bred stock car!
You'll thrill to its performance, you'll hold your breath behind the wheel!!
Every minute is something new never seen before!!!

SEE 306 Ford horses packed into the wild and wooly 289 engine...four-barrel carburation, high rise aluminum manifold and a hand-built tuned exhaust system. **SEE** distinctive Mustang styling to which has been added such "everybody will notice" features as rear quarter panel windows, rear brake scoops and "there goes a G.T. 350" striping. **SEE** computer designed suspension with front anti-roll bar, competition shocks, front disc brakes and torque controlled rear axle. **SEE** the G.T. 350 scream from zero to sixty in 5.7 seconds. **SEE** the G.T. 350...the car that's not for everybody. Just for you.

FIRST	FIRST	FIRST	FIRST	FIRST	FIRST
in Bp at Cumberland. Bob Johnson	overall at Kent, Washington. Jerry Titus	in Bp at Elkhart Lakes. Tom Yeager	overall at Mid-Ohio. Bob Johnson	second, third and fourth in Bp at Lime Rock. Johnson, Donahue, Krinner and Owens	in Bp at Willow Springs. Chuck Cantwell

WRITTEN AND DIRECTED BY **CARROLL SHELBY** PRODUCED BY **SHELBY-AMERICAN, INC.**

COBRA WINS
Le Championnat des Constructeurs

Shelby-American brings the U.S. an international championship

BY JAMES T. CROW

As TIME PASSES, it becomes too easy to forget just how much the U.S. road racing enthusiast and U.S. prestige in international racing owes to Carroll Shelby. Before Shelby undertook the mating of the AC chassis with the medium size Ford V-8 engine (just for the history book, the first Cobra was delivered to the U.S. in February 1962), America didn't have a road racing machine that was enjoying any more than local success. The Corvette was dominating amateur production category racing, of course, but it was falling with a great hollow thump whenever it was matched against the best of international competition. At Sebring, for example, the antics of the Corvettes were positively embarrassing.

The Cobra changed all that. Almost from the very first race, the Cobra put the Corvette to the mat. And since then, Cobra has become a name respected throughout the racing world. Cobras suffered their humiliations, as do all cars that are being seriously campaigned, but Shelby-American stayed at it and in their third year of existence have succeeded in capturing the most highly prized crown to which any manufacturer of Grand Touring cars can aspire, the International Manufacturers' Championship.

Just to keep the record straight, what Shelby-American has won is Division III of the *Championnat des Constructeurs*, to give it the official title assigned by the Federation Internationale de l'Automobile. There are two other divisions in the championship for smaller displacement cars, but almost no one except the manufacturers themselves (and their captive owner clubs) pay any attention to them. The championship was formerly limited to sports cars but since the FIA de-emphasis of sports cars in 1961, the crown has been assigned to Grand Touring cars conforming to the FIA's Appendix J. Ferrari dominated the championship when it went to sports cars (with occasional interruptions by British builders) and continued to win after it went to GT cars in 1962.

THE FIRST RACE in which a Cobra participated came in October 1962, when Bill Krause drove number 0002 in a 3-hr preliminary race before the 1962 Times Grand Prix. The new car didn't run very long, but it made a big impression as Krause overhauled the quickest of the Southern California Corvettes (driven by Dave MacDonald, who was later to become a Cobra star, incidentally) and pulled away into an easy lead until sidelined with a broken rear hub. One of the first of the 1963 Sting Rays won that race, by the way, but it was almost the last event of any importance in which the Corvette was able to score.

The next major effort came the following February at the Daytona Continental and this marked the real beginning of serious campaigning for the new car. The cars failed at Daytona, then failed again at Sebring, both times after showing a turn of speed that made their mechanical failures all the more heartbreaking. But Shelby and his organization were going after success in the only way that success is achieved by a competition car—racing.

Throughout the 1963 season, the Cobras were raced into shape. The cars broke, were stuck back together and raced again. And raced and raced and raced. By the time the season

Cobra Daytona coupes were built at Shelby plant in early 1964.

was over the cars were almost monotonously dependable. They looked ratty, what with the dull paint and the crude metal work, but they were great crowd pleasers wherever they appeared and their contribution toward making the Sports Car Club of America's then-new U.S. Road Racing Championship series a success cannot be overestimated.

During that 1963 season there was no serious effort made toward winning the international manufacturers' championship. Shelby-American ran the cars at Daytona and Sebring (unsuccessfully) and two roadsters with hardtops were entered at Le Mans through AC in England. One of these, driven by Bolton/Sanderson, finished seventh overall and was fourth in the GT category behind three Ferrari GTOs.

THE ASSAULT on the international championship was undertaken seriously in 1964. The first step was to prepare a car that would offer the maximum opportunity to wrest the laurels from Ferrari. A peculiarity in the FIA regulations permitted the construction of special bodies on the basic chassis and running gear of the already-approved model. It was therefore decided that a coupe body would be built for the Cobra roadster, one that would be aerodynamically more efficient and thus better suited for the comparatively higher speeds attained on the international circuits where the championship would be decided.

The coupe body was designed by a Shelby-American employee, Pete Brock. Brock's background includes Los Angeles Art Center school, work in the styling department at General Motors and considerable racing experience in 1100-cc sports/racing cars. The design and construction of the coupe was an exercise in direct and practical body building. Pete worked directly from the chassis and running gear components on which the body would have to fit, made up the wooden body buck and had the aluminum panels rolled to fit these contours at California Metal Shapers. The panels were then fitted to the chassis and the first example completed in time for the Daytona Continental 1000-km race in February. After 200 laps of the 327-lap race, the coupe was more than three laps ahead of its nearest challenger, only to be eliminated by an overheated rear end that led to a fire in the pits.

The message was clear, however, that Shelby-American had a winner in the Daytona coupe and this was demonstrated unmistakably the next month at Sebring. In the classic 12-hr race, again driven by Bob Holbert and Dave MacDonald, the coupe led a 1-2-3 Cobra parade in the GT class and finished fourth overall behind three Ferrari prototypes and a full 40 miles ahead of the first Ferrari GTO. It was after this race that Shelby said he believed that Shelby-American owed it to the sport to seriously attempt to wrest the manufacturers' championship away from Ferrari.

The bright spot of the 1964 season, so far as the Cobra team was concerned in international racing, was the fourth overall at Le Mans scored by the Daytona

CONTINUED ON PAGE 72

Air flow was studied by photographing action of yarn tufts taped to body.

CONTINUED FROM PAGE 18

went off course and bent it front, back and middle. A little later, while Schlesser was still congratulating himself at escaping without serious injury, Britisher Roy Salvadori, driving the other car, planed off the circuit in the rain, got it pointed rear first and bent it so thoroughly that it could not be operated either. Thus, with much distortion of tender metal, the Ford GTs lost the opportunity of learning from experience on the fast Sarthe circuit. In the future, rebuilt and with the best fervently hoped for, two Ford GTs will appear at the Nurburg Ring 1000-km race before the all-out effort that is being planned for the 24-hr race at Le Mans.

Looking at the cutaway details of the Ford GT (above), it seems to this observer that this is one of those interesting machines that has great potential—but that this potential is not going to be realized immediately. It's not a "radical" design, admittedly, but it is a new car and one that should be allowed to be raced into shape before being asked to show itself at such a race as the Le Mans 24-hr.

—*Jean du Pont*

FORD GT MARK II

A 7-liter weapon aimed at winning the Sports-Prototype championship in 1966

BY RON WAKEFIELD

Although the "all-new" Ford GTP sports car prototype pictured in last month's "Miscellaneous Ramblings" will be fielded by FoMoCo at Le Mans, the 7-liter Ford GT in Mark II form will constitute Ford's main prototype effort in 1966. Several significant changes are evident in the new version of the GT and there are myriad small changes of a development nature, most of which are aimed at improving the car's durability inasmuch as there's really no question about its being fast enough. Development work is being done primarily by Shelby American and carried out under the supervision of acknowledged development expert Phil Remington.

Most obvious of the changes are the new body contours front and rear. The new nose is essentially the same as the "production" GT-40, which means that the Mark II is about 9 in. shorter than the previous 7-liter car. The rear end is also new and Ford hopes the various fins and spoilers cobbled up for use at Le Mans last year won't be necessary now. Tests at the Ford proving ground track at Kingman, Ariz., have indicated good stability at 200 mph.

A few pounds have been saved here and there in the body panels by the use of thinner material and there are many structural changes. For instance, there are new jack pads at the front so the same jack can be used front and rear; reinforcing has been added around suspension mounting points at the front; the rear control arm pivot-point studs are now supported on both ends rather than being cantilevered—a small change but apparently an important one. Suspension uprights at both front and rear have been modified for more strength and the new double-adjusting Koni shocks described in February R&T (page 16) are used.

It is fairly well known that the vented rotors used for the disc brakes last year were cracking regularly. To eliminate this problem, the Shelby people have designed a new rotor with the internal ribs curved instead of radially straight. The rotors are also finished by a new process developed by Kelsey-Hayes called "Die Pac"—a coating of 98% copper and 2% carbide which is claimed to give a better coefficient of friction and better heat transfer.

The gearbox, always a problem with this car, now has a single oil pump instead of the double ones used before and is now being made by the Transmission and Chassis Division of Ford rather than Kar Kraft. The engine is pretty much the same as last year, with modifications made to the dry sump lubrication system for better de-aeration of oil returning to the supply tank and a stronger sump for better support of the bell housing. The radiator is larger—now 11.5 x 25.0 x 4.0 in.—and has a new header tank, mounted higher on the front bulkhead than before and equipped with a 21-psi cap.

Engine output will be 460–470 bhp at 6500 rpm and the cars are expected to weigh about 2475 lb full of fuel and ready to race. The Shelby team will campaign these Mark IIs and they will be backed up by a team of lightweight 4.7-liter GT-40s managed by Alan Mann, and the GTP, which will probably be run by Roy Lunn.

THE CHALLENGE TO FERRARI

Porsche Carrera 6/Ford GT Mark II

PHOTOS ALICE BIXLER

THE INTERNATIONAL CHAMPIONSHIP that is being battled over by the greatest names in road racing is this year known as the "International Trophies for Prototype Sports Cars." This is the big one, the one that Ferrari, Ford and Porsche will be contesting in 1966.

The regulations for this championship specify that there will be two displacement classes— over and under 2 liters. Last year Ferrari won the over-2-liter championship, humiliating Ford at Le Mans and winning where it was necessary at the other championship races. In the under-2-liter class, Porsche had it pretty much its own way.

This year, however, Ferrari is preparing a serious challenger for the smaller division as well (see page ←) and to meet this threat, Porsche has introduced its new Carrera 6. This car is all new, not a development of the Porsche 904 which will now run in the Sports 50 class for the other major championship, the International Championship of Sports Car Manufacturers.

The Carrera 6 has a space type tubular frame, fiberglass body and gull-wing-style doors. Typical of Porsche's appreciation of the necessity for driver comfort in long-distance events, the driving compartment is comfortably outfitted and both the steering wheel and the seat are adjustable.

The Carrera 6 has a flat-6 aircooled engine with a bore and stroke of 66 x 80 mm, for a total displacement of 1991 cc. More than 230 bhp is being developed by this engine (the factory gives the figure as 210 DIN hp at 8000 rpm) and there is usable power available between about 3500 and 8200 rpm, a usefully wide power band. A 5-speed gearbox is used and the top speed of the Carrera 6 is given by the manufacturer as 280 kph, roughly 175 mph. The weight of the car without fuel but with oil is about 1275 lb.

The Carrera 6 was very impressive in its first competition appearance, the Daytona 24-hr race. Driven by factory pros Hans Herrmann and Herbert Linge, the Carrera 6 won the 2-liter Sports-Prototype class and finished 6th overall. At Daytona, however, there were none of the new Ferrari Dino 206/S on hand and when these meet in future races, the struggle should be a classic one.

In the big-engined class, Ferrari's number one challenger is the Ford GT Mark II. This has been developed from the 427-cu-in Ford GT that was rushed to completion for last year's Le Mans race and failed after leading impressively

37

THE CHALLENGE TO FERRARI

during the opening hours. The Mark II, powered by the huge 7-liter pushrod Ford engine, has been de-tuned slightly to give "only" about 465-475 bhp. This power, combined with the aerodynamic shape, should propel it along Mulsanne straight at Le Mans at something over 200 mph. Last year's version, less clean aerodynamically, was clocked at 185 consistently and was reported to have touched 210 on occasion. The 427 GTs went out with gearbox trouble last year but this component has been thoroughly tested in the interim and should be ready to go the distance.

There will be a total of eight events counting toward the Sports Prototype championship in 1966. By definition of the Federation Internationale de l'Automobile, races counting toward the over-2-liter championship can be no shorter than 1000 km (621 mi) long or 6 hours duration. Under-2-liter championship races can be half that distance, but as most of the races are for both divisions, the smaller engined cars will be covering roughly the same distance as their larger-engined counterparts.

Counting toward the championships in 1966 are the following races:

Feb. 5-6 24-hr Daytona Continental
March 26 12-hr race, Sebring, Fla.
April 25 1000 kms, Monza Italy
May 8 Targa Florio, Sicily
May 22 Spa GP, Belgium
June 5 1000-km race, Nurburgring, Germany
June 18-19 24-hr race, Le Mans, France
Aug. 14 Hockenheim GP, Germany (2-liter division)
Oct. 2 12-hr race, Reims, France

It is not yet known which of the European races, besides Le Mans, will be entered by Ford. Having already won at Daytona, and expected to make a strong show at Sebring, Ford will have to run (and win) at least three European races in addition to Le Mans if it expects to wrest the championship away from Ferrari. There is little doubt, however, that there is more publicity benefit to be gained by a victory at Le Mans than by backing into the championship by piling up points from lesser wins and there is no doubt at all that Ford's major effort in 1966 will be to win at Le Mans.

ILLUSTRATIONS BY WERNER BÜHRER

AIR INTAKE - ENGINE COOLING

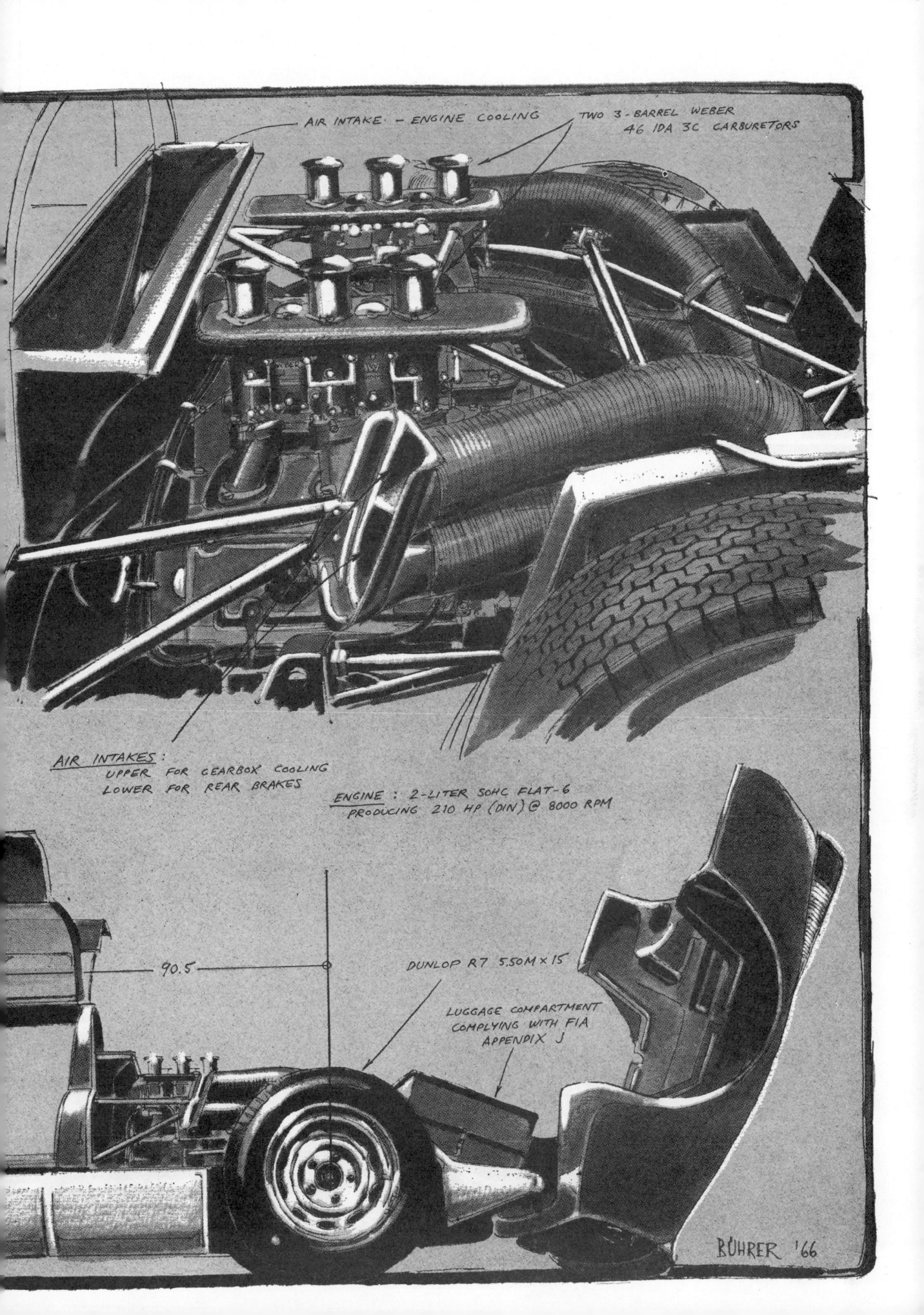

TWO 3-BARREL WEBER
46 IDA 3C CARBURETORS

AIR INTAKES:
UPPER FOR GEARBOX COOLING
LOWER FOR REAR BRAKES

ENGINE: 2-LITER SOHC FLAT-6
PRODUCING 210 HP (DIN) @ 8000 RPM

90.5

DUNLOP R7 5.50M x 15

LUGGAGE COMPARTMENT
COMPLYING WITH FIA
APPENDIX J

BÜHRER '66

FORD 427 GT - MK II

Revised radiator air outlets

40.5

Revised engine air intakes

95.0

Nose shortened 9" from 1965 427

Tires (Goodyear): 9.75 x 15 front
12.80 x 15 rear

1965 LE MANS 427

LONGITUDINAL STABILIZING FINS

LUGGAGE COMPARTMENT COMPLYING WITH FIA APPENDIX J

1965 LE MANS 427

PLASTIC DUCT FROM AIRBOX TO SINGLE HOLLEY CARBURETOR

OIL RADIATOR

BÜHRER '66

TECHNICAL ANALYSIS
PORTRAIT OF THE LE MANS WINNER

BY RON WAKEFIELD
PHOTOS BY SCOTT MALCOLM

THERE'S A LOT of cheer in the Ford racing organization, from Ford general manager Don Frey right on out to the errand boys in the three racing teams. After three years of trying, Ford became the first outfit to win the Le Mans 24-hr race with an American car. And win they did, in grand style with a 1-2-3, even if the finishing order of the three wasn't quite what Mr. Beebe had planned.

Why was the Ford performance so much better this year than in the two years past? Why did Ford campaign the big Mark II, rather than the lighter GT-40 with the 289 engine or the more advanced "J" car? Why a slow-turning 7-liter engine instead of a lighter, high-output unit such as the Indianapolis dohc engine? Or even the 7-liter single overhead cam?

Toward the end of 1964, Roy Lunn, Ford's chief design engineer on the GT project, began to have doubts about the development possibilities of the basic 4.7-liter GT-40. About that time Ford set up Kar Kraft—a wholly owned subsidiary intended to be small enough to get things done with dispatch appropriate to racing ways but close enough to the parent company to draw on its resources.

With Kar Kraft set up and Lunn its head, the first project of the small company was to design a new transmission. The original Colotti had been modified in 20 ways to cope with the relatively large 4.7 engine but its reliability was still marginal; Lunn felt it had to be replaced as transmission failures were still the most common difficulty.

Kar Kraft's next project was to chop up two GT-40 chassis to accommodate the hulking 7-liter Ford Galaxie engine.

Work on these two cars was started in March 1965 with no real thought of running them at Le Mans that year. But tire developments made decent handling possible and with the first running car Ken Miles lapped Ford's 5-mi oval track at Romeo, Mich., at 201 mph. Whereupon somebody up high said this is the car we are going to race!

If Ford people had been experienced at Le Mans, they wouldn't have tried it. They missed the April practice altogether with the 427 cars and actually finished building the second one at Le Mans just before the race, but nevertheless started both cars. What happened is well known now. Hasty preparation resulted in a gear that was intended for scrap being put into one gearbox and dirt on a bearing surface of the other gearbox, putting both the new KK transmissions out of commission. However, from the amount of development work that has been done since then, it seems unlikely that the cars would have finished anyway in 1965.

Lunn says he learned one big thing from the 1965 experience: that the big engine, loafing around at 6000 rpm, was the way to go.

The New Racing Approach

JOHN COWLEY, in charge of managing the racing effort from Dearborn, wisely realized the value of intramural competition. Thus he decided to put *three* racing teams on the job of preparing and racing the cars for the 1966 effort. He stayed at the helm of the operation, coordinating the three teams' work and feeding information back and forth so that

THE Ford-Ferrari / Ferrari-Ford DEAL

What happened when Ford sent Don Frey to Italy to negotiate the purchase of the Ferrari company

IT IS FAIRLY well known that in 1963 Ford Motor Co. set out to buy Ferrari. Until now, however, the details of the story, and even official confirmation of the negotiations, have remained obscure. Here's the story of what happened.

In January 1963, Henry Ford II and Lee Iacocca came up with the idea of buying the Ferrari company. What they had in mind, mainly, was to get into international racing—especially the GT variety. They reasoned rightly that if they wanted a quick start it meant buying brains, experience and facilities. And the best example of all these was Ferrari.

About the same time, word somehow got to Ford of Germany that Ferrari was interested in a merger with Ford. Probably it was one of those periods when Ferrari was short of money. But the important fact is that both parties were interested—and apparently independently so.

In April the Ford people made up their minds. Phil Paradise, head of Ford Italiano, was chosen to make the advance. He did this in May, approaching Mr. Ferrari with the ⟶

each team could benefit from the findings of the others. Previously the teams had gone their own way without much central supervision. Cowley and his two aides—Homer Perry and Chuck Folger, both development engineers—pulled the efforts together while still allowing the initiative and competitive spirits of the separate teams to motivate their work. These three men supervised most of the vehicle tests, which required a staggering amount of travel as tests were run at Daytona, Sebring, Kingman (Arizona, Ford's desert proving ground) and Riverside.

Besides the good coordination, bringing in the Holman-Moody outfit was a major factor in this year's organization. Cowley says that there was a bit of skepticism about a stock-car team's ability to do the job; but H-M certainly proved their own ability at it as well as putting Shelby American even more on its toes. Furthermore, at least one highly significant engineering development that came out of Holman-Moody.

In addition to the new team structure, there was simply greater experience throughout the Ford Dearborn group. By this time the Ford people had gained a good feel for racing. There was a Le Mans Committee, made up of top personnel from the participating divisions of the company—(Engine & Foundry, Transmission & Chassis, and Ford Division) meeting regularly to discuss mutual problems and courses of correction. It all added up to a properly concentrated operation focused on one thing—winning at Le Mans. A brief outline of the people, groups and their functions:

—Donald Frey: VP and general manager of Ford Division.
—Leo Beebe: public relations & promotion, director of racing activity.
—Le Mans Committee: coordinate entire Le Mans activity.
—Jack Passino: manager, Special Vehicles Activity (all GT cars).
—John Cowley: race manager.
—Kar Kraft: design and build original prototypes. Headed by Roy Lunn, with Chuck Mountain, Ed Hull and Bob Negstad on staff, and various "moonlighting" Ford engineers.
—Race teams: Shelby American, Holman-Moody, Alan Mann Racing, Ltd.—prepare, develop and race cars.

The Mark II and its Development

AFTER LE MANS 1965, Ford people were sure that they had in the Mark II (as all 7-liter GTs are called) a car fast enough to be competitive in 1966. If it could last.

The Mark II was basically the same car as the original GT, which had been conceived by Eric Broadley in 1962-63 and refined extensively by Lunn's team of designers. The 7-liter engine and various strengthening components required with it added several hundred pounds of weight; it had a longer nose to accommodate more radiator and ducting, and cast alloy wheels rather than the wire ones of the original GT-40s, now called the Mark I. Conventional in layout for a contemporary GT racing car—a midship engine driving the rear wheels, semi-monocoque chassis construction, short-and- ⟶

43

THE Ford-Ferrari / Ferrari-Ford DEAL

idea of forming two companies out of a merger:

FORD-FERRARI: With Ford as the majority stockholder, this company would build and sell the kind of luxurious sports and GT cars Ferrari was already building.

FERRARI-FORD: The racing company. Ferrari would be the majority stockholder and basically in control but Ford would want to make use of publicity and engineering developments from the racing activities. Also, Ford wanted the option to purchase Ferrari's equity in this company upon the Commendatore's death.

This arrangement seemed essentially satisfactory to both sides, so Ford sent over Donald Frey, Ford Division's general manager, with a team that included an assets-determination specialist, a manufacturing expert and two lawyers to begin the official negotiations. The talks began in mid-May.

Frey says that Mr. Ferrari was sincerely interested in making the deal. Frey stayed in Modena, driving out to Maranello each day to work on details with Ferrari. Ferrari himself rarely arrived before 10 AM and it was usually after lunch before anything was accomplished. Enzo Ferrari is one of those night people, with a late metabolism cycle. Thus work often continued until late at night.

There was little difficulty in agreeing on terms for the Ford-Ferrari part of the deal. Though Ferrari takes great pride in his passenger cars, he has always been closer to his racing. He demonstrated this by being relatively amenable about the passenger car business while having great reservations about the arrangements for the racing organization.

The negotiations got as far as even discussing the emblems to be used on the various cars. Ferrari would sketch out possibilities, with combinations of crests and prancing horses and/or the two names on them. Ferrari placed a figure of $16 million on what Ford would purchase, and at the time negotiations ended, the Ford group had arrived at a figure of $10 million. But there was room for bargaining on both sides.

Frey discussed details of possible racing activities with Ferrari. They talked at length about Indianapolis, and Ferrari surprised Frey somewhat by showing a lot of interest in the Indianapolis 500-mi race and even having a couple of engines he had designed with that race in mind.

But when the talk got down to brass tacks about the racing company—who ran what, who got the publicity and so on, Ferrari began to have doubts about the whole thing. For one thing, he wanted Ford to sever its relations with Shelby-American. Ford, on the other hand, felt an obligation to Shelby and this upset Ferrari. He felt, understandably, that there would be a serious conflict here. Another question from Ford, "What if we wanted to campaign GT cars at Le Mans, promoted by Ford?" brought a significant pause.

Ford's lack of interest in Formula 1 was also off-putting to Ferrari. This was during the time of the 1.5-liter formula, remember. Perhaps if the negotiations had taken place later—when the 3-liter formula was looming ahead—this might not have been a problem.

Why was Ferrari interested in merging with Ford? Frey says that Ferrari admired the elder Henry Ford greatly as a person and respected Ford as a company. He envisioned a happy combination of the large, reputable mass-market car builder combined with his own small-volume artistic approach. He was not interested, definitely, in adapting mass production methods to his passenger cars.

Don Frey recalls some amusing incidents during the ten days he spent in Italy at that time. One evening after a very late-starting day, dinner with Ferrari lasted until 1:00 AM. And for a Ford executive whose regular business day starts around 9:00 AM and winds up about 6:30 PM, this was going a bit far.

On another day, Ferrari drove Frey in one of his latest production models to a favorite inn up in the mountains near Maranello—*Il Gatto Verde*, the Green Cat—and on the way back gave a demonstration of the driving verve for which he's well known. He managed to hang the passenger's side out over the edges of the cliffs and to get Frey's side close enough to the solid walls that Don was more than a little impressed. But Frey was determined to be stoic and rode it out

FORD GT MARK II

long-arm independent suspension front and rear—it was unorthodox in but one area—the large, heavy, slow-turning V-8. Lunn and his people, but not necessarily all the racing teams, were convinced that the big engine was the way to go: it was relatively cheap, well proven (in NASCAR racing) and so related to production engines as to offer maximum advertising potential. There wasn't room, by the way, for the more bulky sohc 427. And a complete development program would have been necessary to make the Indy dohc engine suitable for a 24-hr race.

But with the big torquer came new problems. Most directly related were power transmission components, but most perplexing was the problem of stopping a 2800-lb plus machine

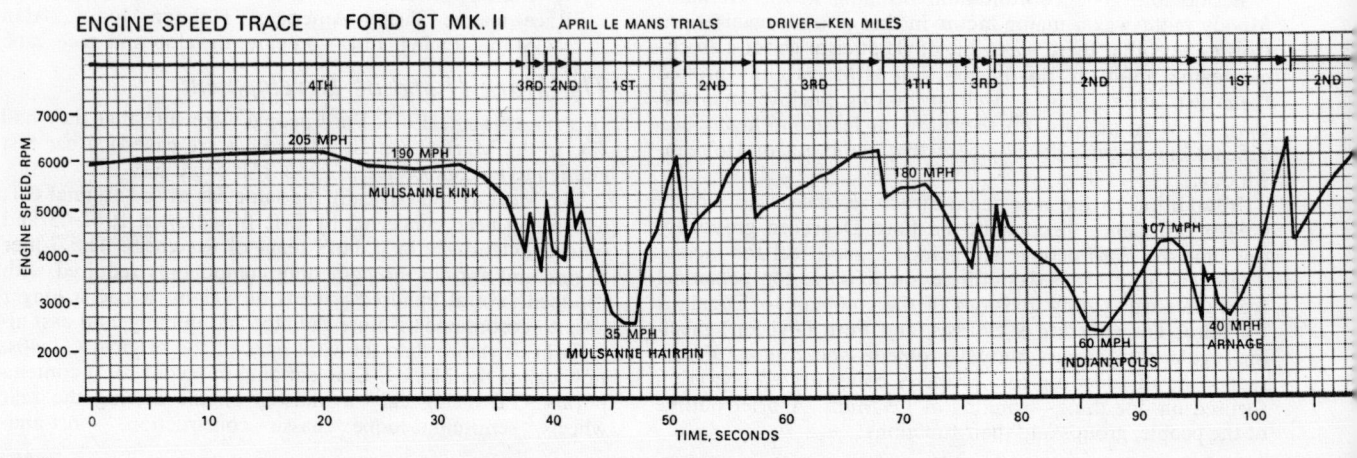

ENGINE SPEED TRACE   FORD GT MK. II   APRIL LE MANS TRIALS   DRIVER—KEN MILES

without a word. Frey also adds that Ferrari's driving was flawless.

Frey was impressed with Ferrari's warmth and volatility. Often described as tyrannical and unapproachable, Ferrari insisted throughout the talks that his loyal employees must be taken care of and was explicit in his proposed arrangements for members of his family.

The talks ended abruptly on Saturday morning after ten days of negotiations. Frey received a phone call from one of Ferrari's lawyers, informing him that there would be no further discussion and that there would be no deal. Disappointed but not really surprised, considering the difficulties the talks had brought to light, Frey and his team packed up.

In June of the same year, Frey journeyed to England where he made the arrangements with John Wyer and Eric Broadley that led to the formation of Ford Advanced Vehicles and subsequently the racing effort that finally led to victory at Le Mans. That's how it all began. —RW

Don Frey, for Ford *Enzo Ferrari, for Ferrari*

(with driver) from 200 mph! The added weight meant harder pounding on suspension pieces at courses like Daytona. The long nose brought on new aerodynamic problems, in the form of poor rear end stability—witness the plethora of fins, tabs and the like on the 1965 Le Mans Mk IIs. However, now there was time to think, to test systematically, to do it right this time . . .

The development program went methodically forward.

A Year of Improvement

ENGINE: After it was apparent that the 7-liter pushrod engine was going to be used, Passino and Shelby asked for all possible weight reduction in the engine. Gus Scussel, engineer at Ford's E & F (Engine & Foundry Division) in charge of the project, achieved it with aluminum heads, an aluminum hub on the vibration damper and a water pump of the same light alloy. This comparison of engine weights is enlightening:

7-liter, NASCAR version.....................602 lb
7-liter, Le Mans version....................550 lb
4.7-liter GT Mk I..........................432 lb
4.2-liter Indianapolis......................428 lb

(all weights dry, less exhaust manifold, air cleaner, clutch)

The nice thing about using this engine, says Scussel, is that he knew at the outset that it had the durability to go 24 hours, providing the rpm range could be controlled. Thus all it needed was refinement, and to keep the speed range where he wanted it, Scussel issued a firm edict to all drivers that 6200 rpm was the limit. No exceptions, even though the engine had a safe limit of 7400 rpm for short-term use. To this end, each car's tach was accurately calibrated and a calibration chart taped in the driver's side door jamb.

Aluminum heads meant a small reduction in valve size from the NASCAR version: gauge diameter of intakes was reduced from 2.16 in. to 2.06, exhausts 1.70 to 1.625. Otherwise head design remained the same, but compression ratio was reduced from 12.5 to 10.5:1 because Le Mans fuel is only 101 octane (research method) vs. the 102.8 allowed in stock car racing. As far as octane requirement is concerned, the 10.5:1 aluminum head is about equivalent to a 10.0:1 iron head because it conducts heat away faster. The regular "hi-riser" intake manifold was retained.

Aside from the aluminum heads, the most important change for GT use was a dry sump. This was completely redesigned this year by E & F and has two scavenge pumps driven by an internal chain from the crankshaft, replacing last year's one pump driven by an external, toothed belt. The pressure pump is gear driven from the camshaft and produces 65-70 psi at 6000 rpm. The oil cooler is a NASCAR item also; cooling is so effective that in cool weather it's necessary to blank off parts of it. Maximum oil temperature under any conditions encountered so far has been 250° F.

Tailoring the engine for the GT also included devising a suitable exhaust system, and carburetion. With the exhaust it was simply a matter of getting the required length of pipe for every cylinder and fitting the resulting bundle into the small space available. Not so simple, after all—there are many (trial and) error bundles lying around the buildup shops!

Somewhat surprisingly, the carburetion settled upon was a single 4-barrel Holley unit, rated at 780 cu ft/min flow. This

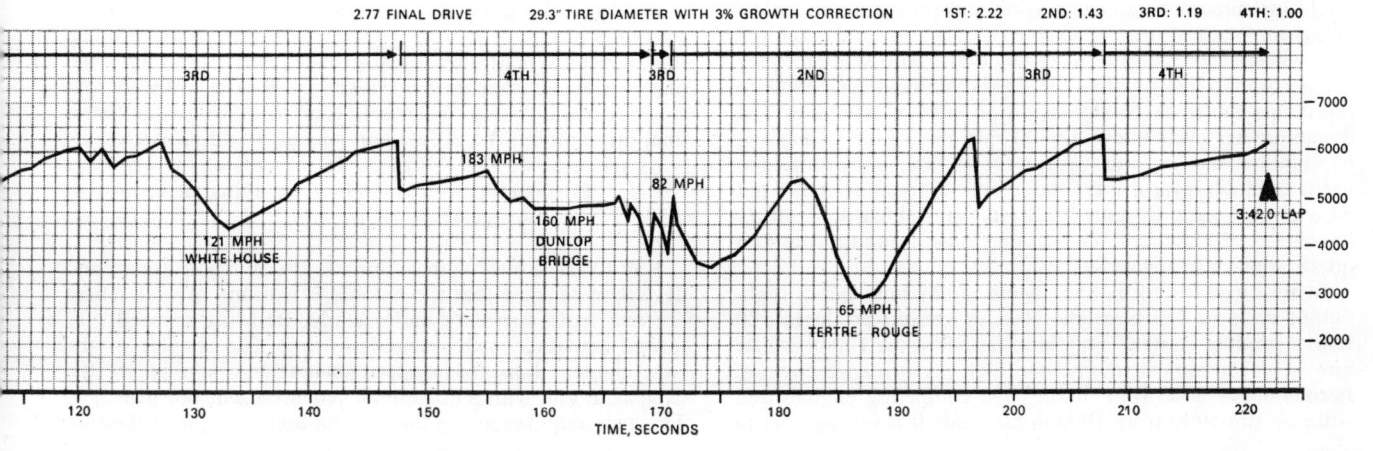

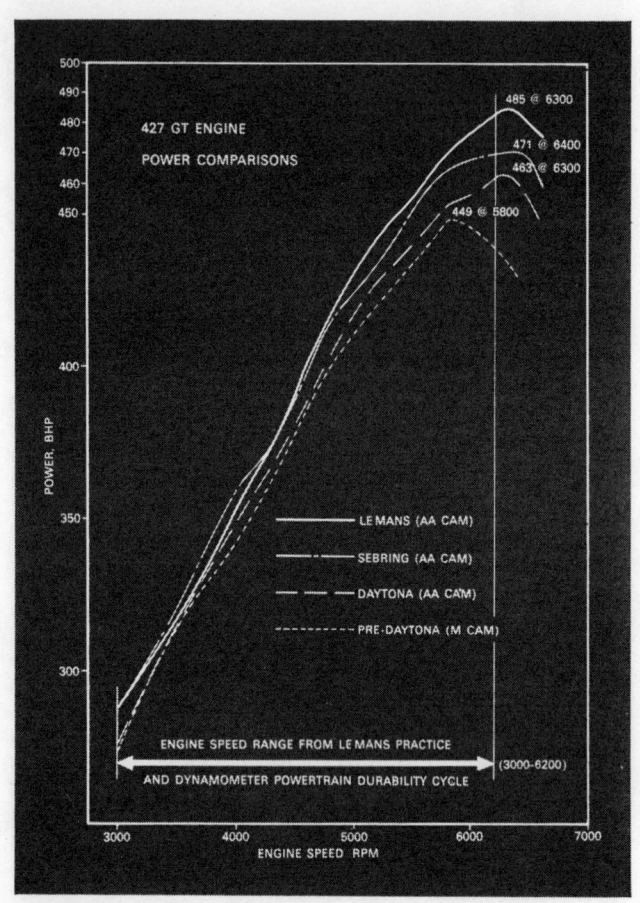

Four power curves for the GT 427 engine show the progress made up to Le Mans time by detail refining of engine.

The 427 GT engine, as installed in the Le Mans cars. Dry sump and exhaust bundle are most obvious features of GT version.

The Holley 4-barrel carburetor with its cam for secondaries.

FORD GT MARK II

looks odd among the multiple Webers of competing cars, but apparently does the job well. E&F wanted vacuum-operated secondary throttles opening, but Ken Miles won out and got mechanical opening. This is accomplished by a cam-and-rod arrangement designed for equal opening, primary and secondary.

Durability testing was thorough in the usual Ford way. Ford dynamometer facilities are extensive and comprehensive, able to accurately duplicate any driving pattern through computer-programmed changes of speed, load, and throttle opening. After preliminary test runs by drivers in an elaborately instrumented car (measuring and recording on an oscillograph such things as engine speed, manifold vacuum, rear wheel speed, throttle plate angle and axle shaft torque) it was comparatively simple for Ford dynamometer people to reproduce the track conditions on an engine test stand. The dynamometer cycle allowed 6800 rpm in 1st and 2nd gears, 6250 in 3rd and top, and was run for 48 hours—as compared to the 6200 limit for drivers and about 38 hours normally put on a given engine (4-hour break-in, 4-hr vehicle sort-out, 6-hr practice, 24-hr race). Thus at race time durability wasn't a question mark in any sense.

In its Le Mans form, the engine is anything but a high-output unit. At 485 bhp (the NASCAR version produces 520 bhp) it produces only 69 bhp/liter, far below the 100 bhp/liter now achieved almost routinely in high performance engines. It runs about 2000 rpm slower than competing engines and will pull smoothly from 1000 in top gear! It is inexpensive to boot, and it is a great achievement to have taken a basic sedan engine and won Le Mans. Complete specifications will be found in accompanying tables.

TRANSMISSION: Getting the tremendous torque of the 427 engine to the ground has been a real challenge for the Ford engineers. However, the problems have been solved and transmission failures are no longer a problem with Ford GTs. Kar Kraft conceived the present transaxle assembly; it is a light-alloy encased unit making maximum use of available heavy-duty Ford gears and shafts.

A Long 2-dry-plate clutch, each plate 10.0 in. in diameter, transmits torque to the input shaft of the gearbox; a production 4-speed synchronized gearset takes over from there, and finally a set of transfer gears takes the torque to the output shaft. It is the transfer gearset that is varied to give different final drive ratios: the differential ring gear and pinion are always the same at 3.09 (34/11). In the case of the Le Mans Mk II, the transfer gearset is 0.899:1 for a final drive ratio of 2.77:1. The four gearbox ratios are 2.22, 1.43, 1.19 and 1.00:1.

Kar Kraft is busy with automatic transmission design and development, now that the manual is working so well. There are two types currently under study. Both are 2-speed plus torque converter with manual shifting. The simpler of the two, lighter than the 4-speed manual, is similar to the Chaparral unit with a dog-clutch synchronized spur gear box. This type requires lifting the throttle foot for upshifts and ju-

Above, front cover assembly, exploded to show chain drive from crankshaft and gears for the two scavenge pumps. Below, front view of sump showing passages and scavenge pickups.

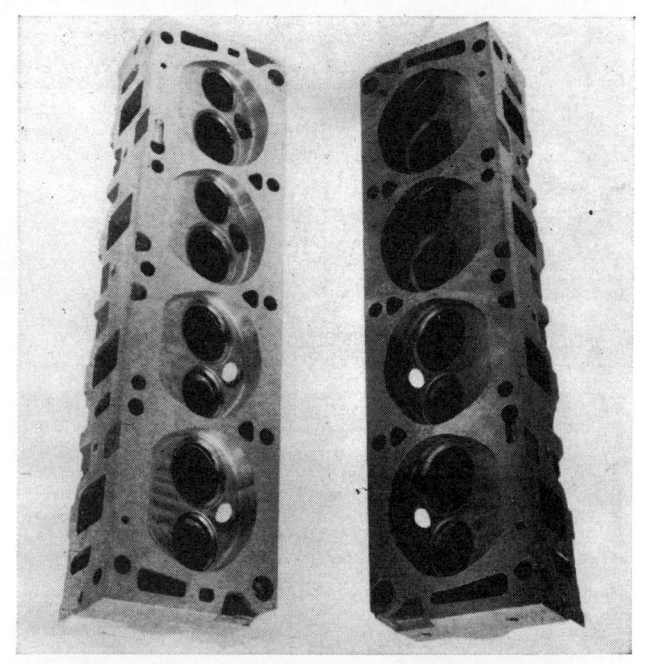

Left, GT aluminum head and right, Galaxie cast iron head.

Drivers used tach calibration charts.

TEST	TACH
800	800
1200	1175
1800	1750
2400	2375
3600	3500
4800	4750
6480	6550

Transaxle has its own oil cooler.

dicious coordination for downshifts—allowing some room for error on the driver's part. The other is a "power-shift" type, which is bulkier and heavier but allows upshifting and downshifting under full power. Again, this is a constant-mesh spur gear box but is shifted hydraulically by oil pressure and disc clutches; the driver's lever operates valves only. It was this unit that was used in one car at Daytona (ran 14 hrs) and Sebring (finished). The torque converter in either transmission is from a Falcon Six, with strengthened vanes, has a stall speed of about 4200 rpm and a maximum multiplication of 1.9:1. For a race like Le Mans, first gear would be about 1.45:1 for an upshift speed of 135 mph. But neither of the automatic boxes is up to Lunn's expectations yet, and development work continues. He is hoping for a maximum power loss of 3-4%, rather low for a torque converter. The typical loss for a modern passenger car automatic, which includes a larger hydraulic circuit for automatic shifting of planetary gearsets, is about 8%.

CHASSIS & SUSPENSION: Changes to the chassis structure have been for convenience and strength. New jacking points at the rear are now the same as those at the front—made possible by the new body shape at the rear. Daytona tests produced cracks in engine mounts and in the structure around the front A-arm pivots, so gussets were added in appropriate places. A-arms up front were increased in diameter, and new front upright castings were designed with greater wall thickness and bearing area, and some internal ribs. The pivot point for the rear upper control arm is now supported at both ends instead of being cantilevered, and the control arm length now can be adjusted without removal. New wheels also resulted from the Daytona tests—they have heavily reinforced spiders. Modifications to date bring the chassis' torsional rigidity to 10,000 lb-ft/degree corrected to the usual 100-in. wheelbase. Most of this development work was carried out by Shelby American.

Roy Lunn says that Ford has stolen the lead in handling and aerodynamics—"If Chaparral and Ferrari had mastered these, we wouldn't have seen them, with their drastically better power/weight ratios!" Again, these achievements have been made possible by Ford's vast engineering resources. For instance, suspension geometry doesn't have to be plotted out on paper time after time until the desired combination is obtained; rather, the computer in effect does this work thousands of times faster than human beings can. Aerodynamics have been studied in the Ford wind tunnel, capable of 130 mph, and checked out on Ford proving grounds under 200-mph steady running. Impressive though Chaparral facilities are for a small outfit, they can't be equated with the vast resources of Ford even if GM help is coming in.

Rather than any revolutionary suspension improvement, then, we must say that suspension geometry of the Ford has been refined to the teeth by the fast trial-and-error of the computer, and as usual fine-tuned by the driver's seat-of-pants. Koni double-adjusting (separate adjustment for jounce and rebound) shocks replace the Armstrong units formerly used.

The original Ford GT as it emerged from Ford Styling in early 1964. Lines were clean and smooth but many lessons in aerodynamics loomed ahead. High front end was problem.

The first Mk II, with its greatly extended nose for the additional cooling the 427 engine requires. Cast wheels, air outlets and tail with spoiler were also new.

The latest Mk II, ready to race. New front end brought length back down; revised rear end has smoother taper, new scoops. Only roof scoops were added for Le Mans.

Instrument layout exhibits legibility, a bit of misspelling and a sense of humor.

Rear support for lower rear A-arm is new; arm was formerly cantilevered from front side of pivot point.

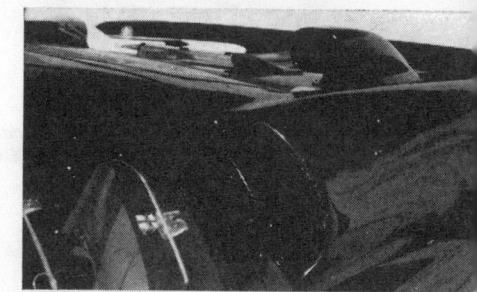

High side scoops are split: upper part to carburetor, lower to brakes. Roof scoops for brakes were added at Sebring.

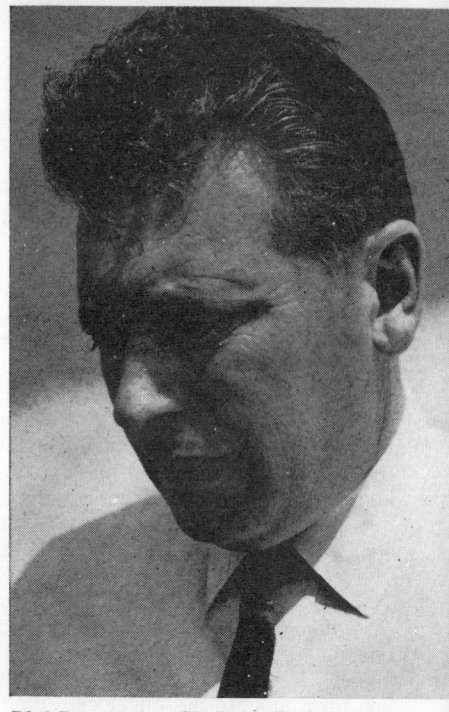

Phil Remington, Shelby's chief engineer.

FORD GT MARK II

Wheelbase remains the same as the Mk I at 95 in., tread is increased 1 in. to 57 at the front and remains 56 in. at the rear.

BRAKES: As noted before, brakes have been a sticky problem. Nobody at Ford or Shelby American will say that the brakes are adequate yet—but rather that the driver must be careful with them. Last year the radial-spoked rotors (vented discs) cracked regularly. Curved spokes this year have reduced failures somewhat but not completely, and metal coatings tried earlier failed to live up to expectations. With 653 sq in. of swept area, there is no room for larger discs or pads. What happens in use is that the rotors get tremendously hot during braking and then cool very rapidly as speed builds.

A Shelby-American mechanic prepares to mate the Kar Kraft transaxle to the engine of the winning car.

Phil Remington demonstrates the ease with which the brake discs are removed. Two bolts hold caliper in place, and wheel itself locates the disc's hat section axially.

FORD GT MARK II SPECIFICATIONS

ENGINE

No. cyl & type . . . . V8, ohv
Bore x stroke, mm . . . . . . 108 x 96
   In. . . . . 4.24 x 3.78
Displacement, cc/sq in . . 6997/427
Compression ratio 10.5:1
Bhp @ rpm . . . . 485 @ 6200
   Equivalent mph 205
Torque @ rpm, lb-ft . . 475 @ 4000
   Equivalent mph 128
Carburetors . . . . . . 1 Holley 780 cfm
   No. barrels, dia. . . . . . . 4 x 1.688
Type fuel required . . . . premium
Lubrication system: dry sump; 1 pressure, 2 scavenge pumps
Ignition system . . . . transistor
Max spark advance . 38° @ 4000 rpm
Alternator capacity, amp 52
Camshaft timing:
   Opening at 0.100 cam lift:
     Intake . . . . 8°30' ATC
     Exhaust . . . . 39°30' BBC
   Closing at 0.100 cam lift:
     Intake . . . . 36°30' ABC
     Exhaust . . . . 11°30' BTC
Camshaft drive . . . . . . silent chain

DRIVE TRAIN

Clutch type . . . . 2-dry plate (Long)
   Diameter, in 10.0
Transaxle: Ford T-44, designed by Kar Kraft, built by Ford T&C Div. Galaxie 4-speed gearset; transfer gears and T&C limited-slip differential; aluminum case.
Gear ratios: 4th (1.00) 2.77:1
   3rd (1.19) 3.30:1
   2nd (1.43) 3.96:1
   1st (2.22) 6.18:1
Synchromesh . . . . on all 4
Differential ratio 3.09:1
Transfer gear ratio 0.899:1

CHASSIS & SUSPENSION

Frame type: semi-monocoque, 0.024–0.049 sheet steel.
Brake type: vented disc, single caliper.
   Swept area, sq in 653
Tire size, front 9.75-15
   Rear 12.80-15
   Make . . . . Goodyear "A"
Steering type . . . . . . rack & pinion
   Ratio 16.0:1
   Turns, lock-to-lock 2.25
   Turning circle, ft 34
Front suspension: independent with unequal length A-arms, coil springs, tube shocks, anti-roll bar.

Rear suspension: independent with trailing arms, unequal length lateral arms, coil springs, tube shocks, anti-roll bar.

ACCOMMODATION

Normal capacity, persons 1
   Occasional capacity 2
Seat width, in . . . . 2 x 15.5
Head room 37.0
Seat back adjustment, deg: variable to driver's preference.
Entrance height, in 39.0
Step-over height 15.6
Door width 33.0

GENERAL

Race weight, lb (tanks full) . . . 2682
   With 150-lb driver 2832
Weight distribution (with driver), front/rear, % 38/62
Wheelbase, in 95.0
Track, front/rear 57.0/56.0
Overall length 163.0
   Width (over scoops) 70.0
   Height 40.5
Frontal area, sq ft 15.8
Ground clearance, in 3.9
Overhang, front/rear 39.0/29.0
Usable luggage space, cu ft 3.5
Fuel tank capacity, gal 42

INSTRUMENTATION

Instruments: 7000 rpm tachometer, oil temperature, oil pressure, water temperature, fuel pressure, ammeter, gearbox oil temperature.
Warning lights: differential oil pressure, engine oil pressure.

CALCULATED DATA

Lb/hp (race weight) 5.8
Mph/1000 rpm (4th gear) 31.5
Engine revs/mi (60 mph) 1905
Piston travel, ft/mi 1200
Rpm @ 2500 ft/min 3965
   Equivalent mph 127
Cu ft/ton mi 166
R&T wear index 21.0

SPEED IN GEARS

4th (6200), mph 205
3rd (6200) 170
2nd (6200) 140
1st (6200) 89

Because pad wear and rotor cracking were insoluble, at least for this year, it seemed that methods had to be found for replacing brake parts quickly. Two neat solutions along this line came forth from the racing teams. Shelby American head Phil Remington devised quick-change brake pad retainers which allow rapid removal and replacement of the pads. A remarkable new feature was conceived by John Holman of Holman-Moody during the year's development: quick-change discs! This is a first, and surely will start a trend. The disc hats are outboard of the hub flange and held in place by the wheel studs, the caliper and the wheel itself. Thus when the wheel is removed, and the caliper swung away (which requires only the loosening of two bolts), the disc may be snatched off and a new one slipped on in seconds. Design details and development work on this item were carried out by Ford and Shelby American engineers. Disc diameters

CONTINUED ON PAGE 71

SHELBY AMERICAN MUSTANGS for 1967

BY RON WAKEFIELD

GT 350 gets the new body, more distinction from Mustang, a rollbar and a lower price— GT 500 with 7 liters added to Shelby line

Our cover car, the new Shelby Mustang GT-500, was to have been the subject of a road test in this issue. But Carroll Shelby's very own optimistic promises couldn't produce an operating example by press time, once again proving that the plans of men are no more certain than those of mice. Just before this was written, Shelby phoned to say that production was finally getting under way, after many delays caused mainly by the unavailability of 1967 Mustang fastbacks from Ford's San Jose, Calif., assembly plant.

The major changes in the Shelby GT body result from the extensively changed 1967 Mustang fastback. This is now really a fastback, as the roofline slopes all the way to the rear of the car. But Shelby's own body modifications are more sweeping than in the past with the result that his cars are considerably more distinguished from the Ford production cars than before.

Grille surround and hood panels are fiberglass and are exclusive for the Shelby cars, and high-beam headlights for the quad light system are mounted centrally in the grille in the manner of rally cars. No chromium decorates the front save that on the bumper and headlight retaining rims, and the only thing keeping bugs out of the radiator is a flat-black expanded metal screen at the back of the grille opening. The simplicity is pleasing.

On the sides, brake scoops are installed where the normal Mustang has simulated ones, and up on the roof quarter panel are interior air-extractor scoops that look like the brake-engine scoops on the Ford Mk II. At the rear, a clever pair of revisions puts a fashionable spoiler onto the deck. The regular Mustang fender caps (separate, bolt-on pieces anyway) are replaced by pieces that kick up and blend into a special deck.

A complete rollbar will be standard in both Shelby Mustangs. An inertia-reel shoulder harness is also new, and detail refinements to front and rear suspension are claimed to improve handling a little. Naturally, spring and shock calibrations are different on the GT 500 to accommodate the extra weight (486 lb of it!) of its larger engine and drive train.

The big gun is the GT 500 with its 428-cu-in. engine. Corresponding to this year's 390 Mustang, the 500 uses a version of Ford's big engine with two 600-cu-ft/min Holley 4-throat carburetors atop a Shelby-developed intake manifold. The 428 has hydraulic lifters designed for high speeds and is not to be confused with the Le Mans-winning 427 engine, though it is from the same family: bore and stroke of the 428 are 4.13 x 3.98 in. vs. the 427's 4.24 x 3.78 in. Advertised power rating for the engine is 355 bhp; Ford claims 345 for its single-4V Thunderbird 428, so somebody's modest. Wonder who?

Transmission choice for the 500 is either the 4-speed manual with 2.32:1 first or a heavy-duty 3-speed automatic, both

SHELBY MUSTANGS

built by Ford. The final drive ratio is 3.50:1 with the manual or 3.25 with the automatic. Wheels are the highly styled Shelby mags, 15 in. x 6.5 in., and have Goodyear extra-low profile nylon tires in the new E 70-15 size.

The GT 350 shares all the new styling features with the 500 but continues with the same 289-cu-in., 306-bhp engine as before. Its transmission choice is the same as for the 500 but axle ratios are 3.89:1 with manual and 3.50 with automatic. Likewise, it will come with a rollbar as standard equipment, and it has the same combination of disc brakes (11.3-in. diameter) front and drums (10.0 in. x 2.5 in.) rear. The only external distinguishing points will be the identification badge and painted designation on the lower front quarter.

A most pleasant surprise on both the new models is price. The GT 350's has been *reduced* from $4428 to $4195, thanks to increased volume of production during the past year, and the new price includes power steering and power brakes. The GT 500's tag is $4395, also including the power assists. Comfort, safety and tractability will be emphasized more in the new models than in the past; thus they will be more suitable for general street use than before, and attractive to a broader range of buyers.

The other car in Shelby's product lineup, the Cobra, continues without changes except that the street version will use a mechanical-lifter 428 rather than a 427 engine. This change actually occurred during the 1966 model run.

1967 SHELBY AMERICAN GT 350 AND GT 500

	GT 350	GT 500
List Price	$4195	$4395
Engine	289 V-8 ohv	428 V-8 ohv
Bore x stroke, mm	102 x 73	105 x 101
Displacement, cc	4736	6989
Bhp @ rpm	306 @ 6000	355 @ 5400
Torque, lb-ft @ rpm	329 @ 4200	420 @ 3200
Carburetion	1 Holley 4-V	2 Holley 4-V
Transmission	4-speed manual or 3-speed automatic	4-speed manual or 3-speed automatic
Final drive ratio	3.89 or 3.50:1	3.50 or 3.25:1
Brakes, front/rear	disc/drum	disc/drum
Tire size	Goodyear E70-15	Goodyear E70-15

	GT 350	GT 500
Frame	unit with body	unit with body
Front suspension: independent with unequal-length A-arms, high coil springs, tube shocks, anti-roll bar (both models).		
Rear suspension: live axle, leaf springs, tube shocks, trailing arms (both models).		
Curb weight, lb	2800	3286
Wheelbase, in	108.0	108.0
Track, front/rear	58.0/58.0	58.0/58.0
Overall length	186.6	186.6
Width	70.9	70.9
Height	49.0	49.0
Frontal area, sq ft	19.2	19.2

SHELBY GT 500

A styled-up fastback that should appeal to a wider audience than previous Shelby Mustangs

ONE THING ABOUT Carroll Shelby and his cars, they aren't subtle. Sneaky, maybe. Sly, perhaps. Deceptive, undoubtedly. But not subtle. His latest entries in the automotive lists—the Shelby Mustang GT 350 and GT 500—are typical. They started with a Mustang fastback, re-styled the extremities, added a rollbar, put in shoulder straps, stuck on some trim and it became a different car. It looks something like a racing car, yet it isn't. It also has something of the flavor of a luxurious Grand Touring machine, but it isn't quite that either. It's different from anything Shelby has offered before—less brutal, less purposefully ugly, less stark performance—and yet it offers an abundance of those virtues for which Shelby's products have become famous. It goes, it handles and it stops.

The appearance is distinctive. At the front the nose has been given a revised shell that does away with the chrome grille above the bumper and adds a scoop below. Where the grille used to be there's now a finished-off snout that is backed by an expanded-metal bug strainer and encases a pair of close-together headlights mounted in the manner made popular on international rally cars. The hood, also fiberglass, has a big bulge molded in behind a functional airscoop and there are a pair of post-and-peg hold-downs.

On the sides of the car, the alterations consist of two pairs of scoops. Two fit over the cockpit air extractors in the rear quarter area; the others offer fresh air to the rear brakes.

At the rear, there's a moderate ducktail on the deck lid. This effect has been achieved by using a fiberglass trunk lid and replacing the standard rear fender caps with ones that have a matching upsweep. Across the back there are two wide tail-lights in place of the triplicated smaller ones on the standard Mustang.

These changes in appearance, plus shiny 15-in. steel wheels and Goodyear E70-15s, hang together well, in our opinion. It looks like what it is, a styled-up version of the Mustang fastback. Shelby would like to have you believe that this sort of thing "just happens" at Shelby American but the sure hand of a thoroughly professional stylist obviously had more than just a little to do with these changes. The original GT 350 looked appropriately purposeful but lacked the class of the current model.

The interior has the same sort of distinctive-but-similar flavor as the exterior. The Mustang seats, instrument panel, controls, etc., are retained but there is the addition of a proper rollbar and there are shoulder straps to supplement the standard lap belts. These shoulder straps are attached to the rollbar through an inertia reel that allows the wearer to lean forward as long as it is done slowly but locks up solid on being yanked.

The basic instrumentation of the Mustang is supplemented by an ammeter and oil pressure gauge awkwardly located on the bottom edge of the middle of the dash. There's also a wood-rimmed steering wheel with the Shelby emblem and this, being less deeply dished than the standard Ford wheel, is consequently a bit farther from the driver's chest. Which is good.

The back seat of the fastback is retained this year and though the seating is minimal, it does make a practical 2+2 where the original GT 350 was strictly a 2-seater. This back seat will also fold flat to make an attractive and practical flat deck and, borrowing an idea from the Plymouth Barracuda, there is a drop-down door between the small trunk and the back seat to make even more space.

The suspension of the GT 500 is stiff and the ride could be described as extra-firm. Until now, Shelby American has lowered the pivot point of the front upper A-arm but since Ford incorporated that change in the 1967 Mustang, this is no longer necessary. Shelby still uses stiffer springs front and rear, a larger anti-roll bar front (0.94 in., not as large as last year's 1.00) and Gabriel adjustable shocks all around.

At the rear, the trailing arms previously added to the GT 350 are gone, replaced by rubber snubbers mounted 8 in. be-

PHOTOS BY HENRY N. MANNEY & GORDON CHITTENDEN

hind the front eyes of the leaf springs. These still provide some resistance to rear axle windup and hop but are not as effective—or as harsh—or as expensive—as the arms. They did the job in our test car (the GT 500 with automatic transmission and without limited slip) but we don't know if they'd do equally well with the manual gearbox, hard clutch and limited-slip differential. The milder suspension alterations this year represent part of Shelby's effort to tailor the cars to a wider market—and at a lower cost.

The brakes are the disc front/drum rear combination available as an option from Ford but with a more fade-resistant organic friction material. These were power assisted on our test car and though the touch is a bit lighter than we prefer, they are comparatively easy to control. Because of the weight of the GT 500, the swept area per ton is not impressive (175 sq in/ton) and by the time we'd completed our sixth stop from 60 during our fade tests, the pedal effort had increased by 48%.

This year's Shelby Mustang GT can be had with the 289-cu-in. V-8 (the GT 350), a supercharged 289, or with the big-big 428 (the GT 500). This 428 (4.13 x 3.98 bore and stroke) is Ford's big cheap cooking engine used in the Thunderbird and the Police Interceptor variants, you should understand, not the celebrated 427 (4.24 x 3.78) developed for NASCAR stock car racing and used in the Le Mans-winning GT Mark II prototypes. As fitted to the GT 500, the 428 has hydraulic valve lifters, is equipped with two Holley 4-throat carburetors and is rated at 355 bhp at 5400 rpm.

In pure physical bulk, the 428 is bigger than king size. There is barely room for it in the Mustang's hull and though there have to be sparkplugs down there someplace (you can see the wires disappear under the rocker covers), changing them doesn't even bear thinking about.

That the 428 is also full of weight as well as bulk is demonstrated by the 3520-lb curb weight of the GT 500. The standard GT 350 we tested earlier had a curb weight of 2800 lb.

Our GT 500 had not only the 428 engine and automatic transmission, power steering and power brakes, but also air conditioning. All these things, with the exception of the air conditioning, come standard at no extra cost on the GT 500. Although these power assists may seem inappropriate to such a car at first glance, they blend into the car's personality without obtrusion. And unless you have biceps like Freddie Lorenzen, you'll find the power steering almost mandatory what with those big tires and all that weight pressing them against the ground.

The car is extremely easy to drive. The engine lights off with a whump, there's a clunk-jump when the shift lever is moved into gear, and if you mash on the gas pedal, you'll GO. The steering is easy, it tends to go where it is pointed, and on the open highway it rolls along without fuss or fury. There is an unmistakable detent in the throttle operation so

---

SHELBY GT 500
AT A GLANCE

Price as tested	$5114
Engine	V-8, ohv, 7014 cc, 355 bhp
Curb weight, lb	3520
Top speed, mph	132
Acceleration, 0-¼ mi, sec	15.5
Average fuel consumption, mpg	9.8

Summary: More refinement and styling distinction than past GT 350, power assists standard; performance of big 7-liter disappointing. The ultimate attention-getter.

55

SHELBY GT 500

it requires a conscious effort to get through to the carburetors' secondaries. When you do push down hard enough to bring in the additional barrels there's a great hollow gasping, gouts of smoke pour out the rear and the car hunches forward with a bellow.

In pure acceleration, however, the GT 500 simply doesn't have anything sensational to offer. As we tested the car (two up, plus test gear), it would do 15.5-sec standing quarters consistently but that was about all. And that simply isn't very fast as dragstrip times go. A Ford Mustang with the 390-cu-in. engine option does as well and based on our experience, it would take about 400 bhp and a stick shift to get the GT 500 down to the 13.5-sec quarter claimed for it in Shelby American publicity.

As for handling, the GT 500 is something less than we've come to expect from Shelby's cars but still very good in comparison to the typical American sedan. On the other hand, considering the weight distribution, it's better than we would have thought possible only a couple years ago. With 58% of the total weight on the front wheels, we'd expect it to have understeer akin to that of the USS United States, but it doesn't. As we said, it goes where it's pointed. With that amount of weight at the front, there's no doubt that the front tires are operating at greater slip angles than the rears when the car is turning but we have a theory that modern widetread tires such as those on the GT 500 are so generous in their grip that non-break away understeer is almost undetectable. And especially when you're even further insulated from what's happening by power steering. If you push the car hard enough, the front end will finally slide but the grip is so good that cornering limits are higher than even normally vigorous driving will ever bring out.

Some of the things we didn't like about the GT 500 included the amount of attention given to it by traffic officers.

We also found our particular example needed the air conditioning working most of the time. Not only was there an uncomfortable amount of heat coming through the firewall, there were strong gasoline vapors that became even more objectionable when the throttle was depressed. With the air conditioner running, the heat was counteracted and the odor of raw gasoline diminished. The test car also deposited a pool of oil every time it was parked.

The GT 500 also recorded the highest fuel consumption of any street version car we've tested in years. In over 800 miles, we averaged 9.8 miles per gallon.

The noise level of the GT 500 was reasonably low and though there seemed to be a lot of valve clatter for an engine with hydraulic lifters, it blended in with the general rumble and was soon forgotten. There was a distinctive noise from the back end, though, that on a slightly rough surface sounded like the trunk was full of roller skates.

All in all, though, the GT 500 is a more civilized vehicle than the original GT 350 from which it descended. It rides better, it has more amenities and it is far more attractive. It isn't so closely related to a racing car, perhaps, but we have the feeling that it will appeal to a larger number of buyers than any previous Shelby American automobile.

Big 428 V-8 engine leaves little extra space under hood.

Trunk is minimal size but there's more space behind rear seat.

ROAD TEST
SHELBY GT 500

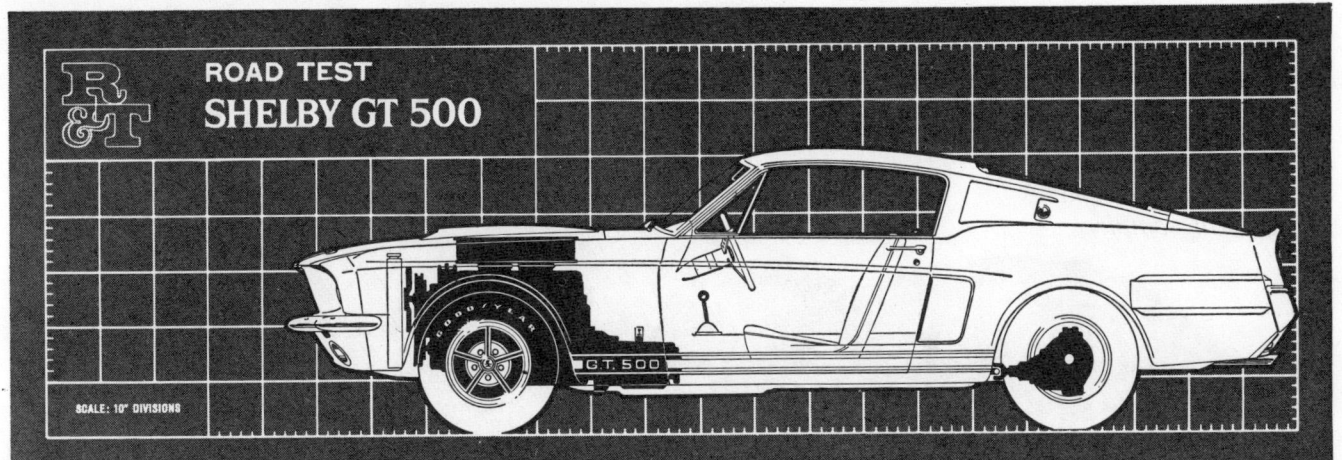

SCALE: 10" DIVISIONS

PRICE
Basic list.................$4395
As tested.................$5114

ENGINE
Type....................V-8, ohv
Bore x stroke, mm......115 x 101
 Equivalent in......4.13 x 3.98
Displacement, cc/cu in...7014/428
Compression ratio..........10.5:1
Bhp @ rpm..........355 @ 5400
 Equivalent mph.............127
Torque @ rpm, lb-ft..420 @ 3200
 Equivalent mph..............75
Carburetion..........2 Holley 4-V
Type fuel required.......premium

DRIVE TRAIN
Automatic transmission: Ford Cruise-O-Matic (torque converter with 3-speed planetary gearbox).
Gear ratios: 3rd (1.00).....3.25:1
 2nd (1.46)..............4.75:1
 1st (2.46)..............8.00:1
 1st (2.46 x 2.1).......16.8:1
Final drive ratio..........3.25:1
Optional ratio.............3.50:1

CHASSIS & BODY
Body/frame unitized with reinforced floor members.
Brake type: 11.3-in. discs front, 10x2.5-in duo-servo drums rear.
 Swept area, sq in..........330
Wheel type & size, in.....15 x 6.5
Tires...Goodyear Speedway E70-15
Steering type: recirculating ball, power assisted.
 Overall ratio............20.3:1
 Turns, lock-to-lock.........3.6
 Turning circle, ft.........39.4
Front suspension: independent with short & long arms, coil springs, tube shocks, anti-roll bar.
Rear suspension: live axle with semi-elliptic leaf springs, anti-windup snubber, tube shocks.

OPTIONAL EQUIPMENT
Included in "as tested" price: air conditioning, automatic transmission, radio, crankcase emission control, fold-down rear seat, styled wheels.
Other: limited-slip differential, exhaust emission control (Calif.), 12/12,000 warranty on powertrain.

ACCOMMODATION
Seating capacity, persons...4 + 1
Seat width, front/rear..2 x 20/40
Head room, front/rear....39/35.5
Seat back adjustment, deg......0
Driver comfort rating (scale of 100):
 Driver 69 in. tall.............95
 Driver 72 in. tall.............90
 Driver 75 in. tall.............80

INSTRUMENTATION
Instruments: 140-mph speedometer, 8000-rpm tachometer, fuel, water temperature, oil pressure, ammeter, clock.
Warning lights: parking brake, seat-belt, high beam, directionals, alternator.

MAINTENANCE
Crankcase capacity, qt..........5
 Change interval, mi........6000
Filter change interval, mi.....6000
Chassis lube interval, mi.....6000
Tire pressures, psi..........28/28

MISCELLANEOUS
Body styles available: fastback as tested.
Warranty period, mo/mi: 3/4000 on powertrain; 12/12,000 on car.

GENERAL
Curb weight, lb............3520
Test weight................3900
Weight distribution (with driver), front/rear, %....58/42
Wheelbase, in..............108.0
Track, front/rear........58.0/58.0
Overall length............186.6
 Width....................70.9
 Height...................49.0
Frontal area, sq ft.........19.3
Ground clearance, in........5.1
Overhang, front/rear....33.2/40.5
Usable trunk space, cu ft.....6.0
Fuel tank capacity, gal......17

CALCULATED DATA
Lb/hp (test wt).............11.0
Mph/1000 rpm (3rd gear)....23.8
Engine revs/mi (60 mph)....2520
Piston travel, ft/mi........1670
Rpm @ 2500 ft/min........3770
 Equivalent mph............89
Cu ft/ton mi...............160
R&T wear index............42.1
Brake swept area
 sq in/ton.................170

ROAD TEST RESULTS

ACCELERATION
Time to distance, sec:
 0–100 ft..................3.4
 0–250 ft..................5.8
 0–500 ft..................8.6
 0–750 ft.................11.0
 0–1000 ft................13.2
 0–1320 ft (¼ mi).........15.5
Speed at end of ¼ mi, mph...95
Time to speed, sec:
 0–30 mph.................2.8
 0–40 mph.................4.1
 0–50 mph.................5.7
 0–60 mph.................7.2
 0–70 mph.................8.8
 0–80 mph................11.0
 0–100 mph...............17.5
Passing exposure time, sec:
 To pass car going 50 mph...4.5

FUEL CONSUMPTION
Normal driving, mph.......8–12
Cruising range, mi......135–205

SPEEDS IN GEARS
3rd gear (5600 rpm), mph....132
2nd (5600)..................90
1st (5600)..................54

BRAKES
Panic stop from 80 mph:
 Deceleration, % g..........84
 Control..................good
Fade test: percent of increase in pedal effort required to maintain 50%-g deceleration rate in six stops from 60 mph..........48
Parking brake: 30% grade....yes
Overall brake rating.......good

SPEEDOMETER ERROR
30 mph indicated......actual 29.0
40 mph...................38.5
60 mph...................57.5
80 mph...................77.0
100 mph..................97.0
Odometer, 10.0 mi...actual 10.16

ACCELERATION & COASTING

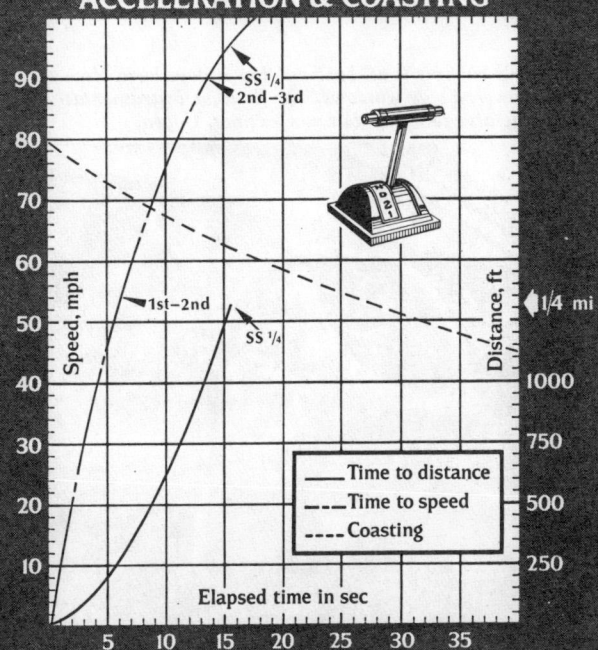

Time to distance
Time to speed
Coasting

57

SHELBY AMERICAN: 1968

Carroll Shelby reshapes his growing empire

The Shelby Lone Star prototype, powered by a midship Ford 289 engine, is a candidate for limited production at about $15,000 the copy.

SCOTT MALCOLM PHOTOS

Lone Star's cockpit is all business, but it does have electrically powered side windows. The familiar instrumentation is from the also familiar, but now extinct, Cobra.

SHELBY AMERICAN has been strangely quiet of late, even though Carroll Shelby dropped by our office some months ago to tell us of the reorganization of his operations. The changes in the structure of SA, and the changes in just what Shelby is doing these days, are somewhat surprising.

Formerly all SA operations were carried on within one company—Shelby American, Inc. As Shelby himself puts it, "We didn't know which hand was feeding us and which was robbing us." GT-350 production might be supporting the racing division, for instance. Now Shelby American is a holding company and it contains four separate companies, each with its own bookkeeping: the racing company, called Shelby Racing Co., Inc.; the accessory company, which markets things like Cobra T-shirts and Pit Stop deodorant; the parts company, which produces Cobra kits for Fords (marketed through Ford dealers) and the Shelby performance accessories, marketed by independent accessory houses; and finally, Shelby Enterprises, including such things as a motel in Tahoe, Calif., Shelby's planned boy's home ranch in Terlingua, Texas, and several other unrelated activities.

The Ford-AC Cobra is out of production, as far as the U.S. is concerned. AC Cars will continue the 289-powered roadsters for their home market for a while, and the tubular chassis developed by AC and Ford for the 427 will still find use in that car and the luxurious AC 428 models. In all, Shelby built 630 of the 289s, 510 of the 427-428s.

Shelby is completely out of series car production, in fact. Though Shelby's design team (in the racing company) decides what the Mustang GT 350 and 500 will be like, Ford Motor Co. is now in charge of engineering and production of the cars. They are now built in a plant at Ionia, Mich. And they carry the name Cobra, you'll notice in the ads. Ford now controls the use of that name.

Shelby has purchased a 4-story office building in Playa del Rey, Calif., from which all the operations are overseen. The racing, parts and accessory companies are housed in a newly purchased 84,000-sq-ft building in Torrance, 10 miles south.

Asked why production of the Cobra was stopped, Shelby is candid: the safety regulations. Meeting them requires too much engineering time, as far as he is concerned. Will there be a new Cobra?—"Only if the government decides to exempt small car makers. If not, I'll quit. But I'm not crying. Production was taking too much of my time anyway." In case there is a chance to build an up-to-date Cobra, Shel has a very modern prototype just about ready. It's called the Lone Star and is shown on these pages.

The Lone Star's specifications are pretty much those of a contemporary road racing car. Built around a tubular chassis similar in construction to the Cobra's, it carries a 289 Ford engine amidships and could carry any derivation of that unit, but not the larger 427. It has a wheelbase of 92.8 in. and a track of 54.0 in. front and rear. Overall dimensions are: length 161.0, width 68.0, height 40.0 in. Its 2040 lb are distributed 44% front, 56% rear. The 289 engine drives through a 5-speed ZF transaxle. Brake and suspension are typical of contemporary practice: Solid disc brakes and unequal-A-arm geometry. Though designed as a race car, it is fitted out luxuriously, with electric side windows and a detachable roof.

Al Dowd and Phil Remington run the racing company. They have four project engineers, all of them with SA before the shakeup. The three main activities right now are Group 7 (the King Cobra), Group 2 (Mustangs for Trans-Am) and the evaluation of the Toyota 2000 for competition work. John Timanus still runs the Driving School within this company, but its scope is not what it was before—the student must furnish his own car.

We asked Carroll Shelby if he had anything he'd particularly like to say, to get off his chest. Leaning back in his plush office, with western music playing softly on the large stereo, Shel said he didn't have much to say, that he was happy and felt he was doing pretty well "for an old driver." But he did allow that he wished auto racing could be half as well organized and sanctioned as football is . . . which seems like a good idea to us, too.

Carroll Smith, Phil Remington and Al Dowd (left to right) pore over a design detail on the King Cobra Group 7 car at the shop in Torrance, Calif. Only one of the King Cobras is running presently so that sorting-out efforts can be focused.

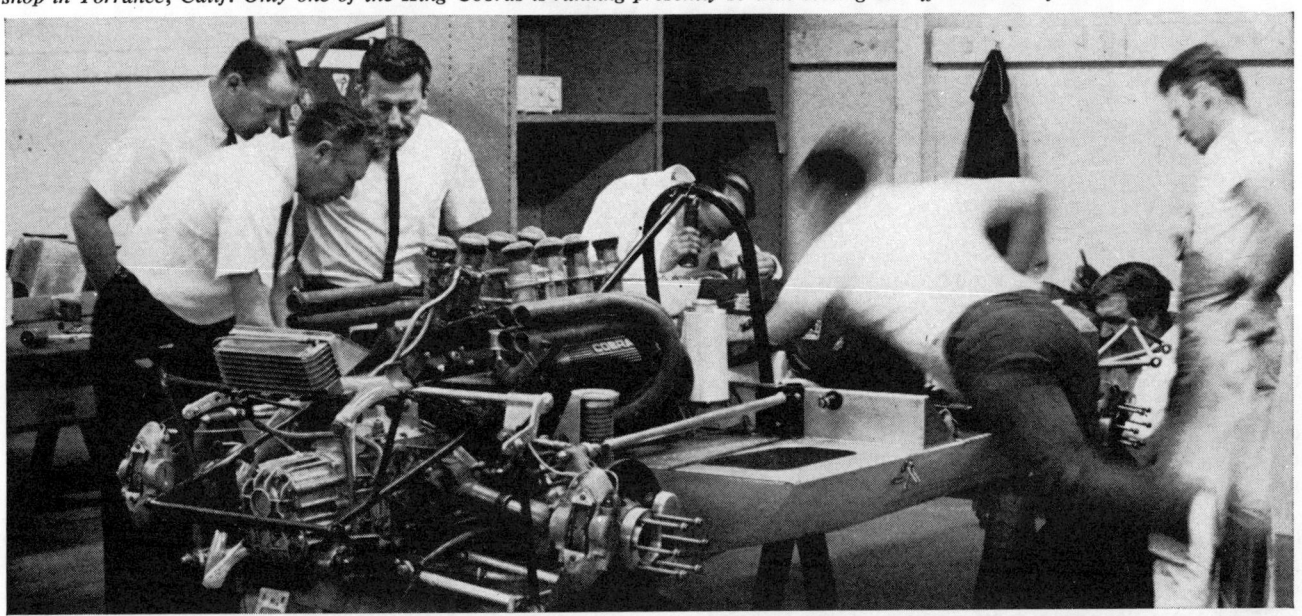

One current activity is the shaping-up of the Toyota 2000 GT for racing. This body shell is being studied for lightening.

In the dynamometer room the Toyota's engine is evaluated and developed. All this work is purely a short-term contract.

Carroll Shelby's Personal Cobra is the
COBRA TO END ALL COBRAS

WE CALLED the 1968 Corvette psychedelic. But that was before we saw Carroll Shelby's new and personal Cobra. All the other mad cars that have passed through the portals of R&T—the Fiberfab Avenger, the Reliant Regal, the Excalibur SS, the Fiat-Abarth OT 1600, the Cyclops, Shelby's own Mustangs and Cobras of past years—all of them pale by comparison. Shel's Cobra is not only the Cobra to end all Cobras—it's the last of the series—but it's the Cobra for the man who has everything. And who's to pretend that Shel doesn't have everything?

So what's different about Carroll Shelby's Cobra? First, last, and most, two—count 'em, two—whopping Paxton superchargers atop the already potent (about 450 bhp) 427 engine. Each of them feeding a huge Holley 4V carburetor. Each of them delivering 6 psi pressure to the carburetor. How much power? Shelby says 800 bhp. That might be stretching it a bit. Perhaps one should just fall back on the old Rolls-Royce answer and say, "adequate."

Behind this incredible engine, everything is pretty normal; just a few little touches to complete the picture properly. A beefy Ford T6 3-speed automatic transmission drives through a 3.31:1 limited-slip differential and the standard Cobra independent rear suspension to 12.10-15 Goodyear tires on 9.5-in. wheels. With no clutch to deliver shock loads to the rear wheels, the big tires have no trouble at all putting the tremendous torque to the ground.

In the cockpit is a marvelous array of instruments, enough of them to warm the heart of any pureblooded enthusiast. In front of the driver are the essentials demanded by any race driver: tachometer, oil pressure and water temperature gauges. The tach reads to 8000 rpm. Out in the center array are a 180-mph speedometer, ammeter, fuel pressure and oil pressure gauges, blower pressure gauges, and intake manifold vacuum gauges that read also into the positive side of the scale when the blowers are going full tilt. Blower pressure and manifold pressure readings are nearly identical under full pressure conditions. Two auxiliary electric fuel pumps supplement the engine's mechanical pump when the extra supply is needed. And believe us, it'll be needed if the performance potential is used, for the gentlest driving nets just about 9 mpg!

Not wishing to hide his candle under a bushel, Shel had his metalbenders dress up the Cobra a bit here and there. A giant hood bubble clears the two blowers and forms a big airscoop at the front. The exhaust headers emerge chromed from the front fender and blend into a large collector/small muffler on each side before giving their all to unsuspecting bystanders just ahead of the rear wheels. Our photographer Scott Malcolm still has a raw red calf from *that*. A chrome rollbar and competition seat belts furnish crash protection. The delicate aluminum bodywork has been bent, smoothed and painted beautifully (a dark metallic blue) by Shelby's body shop, but the finish won't last long if the rear tires keep scraping their fenders and hapless road testers continue to get caught on roads with bits of gravel strewn about. Those latter particles all leave small outward dings in the aluminum surface, as there are no inner fender panels. The car weighs 2500 lb, full of lubricants and water and with 10 gallons fuel.

With a double-blown 7-liter and three fuel pumps to make demands upon it, the fuel tank is adequate for some traveling at 42 gallons. That's the regular competition Cobra capacity.

By this time the reader just might be wondering what it's like to drive a machine like this. Well, there's one thing above all: it's *conspicuous*. Besides simply looking like the meanest thing short of a Group 7, it also sounds that way. The engine does idle at about 700 with the transmission engaged but not without audible protest. And holding the car against the torque converter goes from *tough* with the engine idling to *well-nigh impossible* with it revved up. This latter condition forced us to make simple, drive-off starts in our ac-

Monstrous hood bubble clears blowers and gathers fresh air. Notice the ultra-smooth bodywork and beautiful paintwork.

Ford T-bar selects ranges for the automatic transmission. All four small instruments on right deal with superchargers.

END-ALL COBRA

celeration tests. Not only did the engine tend to load up when we attempted to get it up to converter stall speed, but we physically couldn't hold the car still with the brakes! The brakes are vacuum assisted, but—wow!

Anyhow, Shel's Cobra may not be exactly designed for ambling about town, but that's what it winds up doing most of the time and it does it well, everything considered. The alternator-water pump belt came off while we had the car, but prior to that it showed no tendency to overheat in traffic. The engine makes a lot of fuss at idle, but it's always ready to move flexibly off from the traffic lights.

But the fun comes when you can put your *FOOT IN IT*. And when you do, prepare to be impressed and compressed!

We took it to our favorite dragstrip, Orange County Raceway, for the acceleration tests. With OCR's electronic timing in action we got readings of 11.86 sec for the standing ¼-mi with a final speed of 115.5 mph. We let the automatic shift for itself at 6200 rpm, as there was nothing to be gained by taping it on to the allowable 7000 rpm. Now, that may not sound startling. But, mind you, that was achieved *without wheelspin*! The blue Cobra doesn't get off the line like a cannon shot, but when those blowers start whining . . .

Top speed? We didn't run it. The Cobra isn't an aerodynamic wonder, but 182 @ 7000 rpm seems reasonable with the output and the gearing. By the way, a duplicate of the car has been built for entertainer Bill Cosby. So you're twice as likely to see one of these monsters roaming about. Beware.

As mentioned last month, the "classic" Cobra is now out of production in the U.S. The Shelby and Cosby cars are the final two examples to be built so this machine is, literally, the Cobra to end all Cobras.

It's madness . . . sanitary installation of two Paxton blowers tops off a chrome-bedecked 427 engine for adequate power.

ACCELERATION

Time to distance, sec:
- 0–100 ft....................2.6
- 0–250 ft....................4.5
- 0–500 ft....................7.0
- 0–750 ft....................9.0
- 0–1000 ft..................10.6
- 0–1320 ft (¼ mi).........11.9

Speed at end, mph..........116

Time to speed, sec:
- 0–30 mph..................2.2
- 0–40 mph..................2.6
- 0–60 mph..................3.8
- 0–80 mph..................5.6
- 0–100 mph.................7.9
- 0–120 mph................13.6
- 0–140 mph................24.4

Passing exposure time, sec:
- To pass car going 50.........2.1

SPEEDS IN GEARS

- 3rd gear (7000 rpm), mph......182
- 2nd (7000)..............126
- 1st (7000)...............75

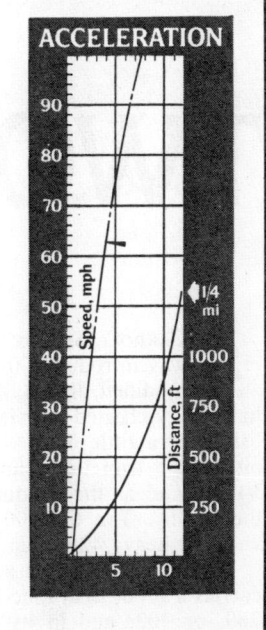

TWO SHELBY GT 350s

Shelby Mustangs, cool and hot

CARROLL SHELBY'S GT 350 has come a long way. It was introduced in 1965 and started out as a much-modified, light (2800 lb), hot (306 bhp from 289 cu in.), hairy-chested version of the Mustang fastback and listed for a little over $4500. In 1967 Shelby backed off a bit on the trim items and lowered the price to more like $4100, and as the production Mustang grew in size so did the Shelby. The GT 500 version came along in 1967 too, with the heavy 428-cu-in. engine dropped in.

In 1968 the cycle is almost complete. The GT 350 now carries a mild, hydraulic-lifter version of the newly enlarged 289, or 302, and in its standard form is almost undistinguishable from the regular production Mustang 302 in performance and handling. No longer are alloy wheels standard, and last year's standard inertia-reel safety harnesses are now optional. The GT 350 now weighs over 3300 lb, requires power steering for parking and, in basic form, should be considered a Mustang with a racy-looking trim package. For really hot performance one must move on, past the hydraulic-lifter 428 GT 500 to the full-house 427 version.

All the external identifying marks are still there, to be sure. The front end is extended many inches by the distinctive fiberglass hood and frontispiece; the hood is full of scoops and vents, both real and imagined; the sides are adorned with false scoops; and the rear end has a "spoiler"

TWO SHELBY GT 350s AT A GLANCE

Price as tested............................$5368
Engine.................V-8, ohv, 4949 cc, 315 bhp
Curb weight, lb..........................3335
Top speed, mph..........................119
Acceleration, 0-¼ mi, sec................14.9
Average fuel consumption, mpg...........13.5
Summary: high-output hydraulic lifter engine is strong & quiet but very expensive . . . good emergency braking but lots of fade . . . high hood makes traffic maneuvering difficult . . . the best passenger restraint system available for production car.

TWO SHELBY GT 350s

in the Detroit sense. But the original idea of the GT 350 —that of making the Mustang as nearly as possible a sports car—has clearly faded into the dim past. And if it can be argued that the GT 500s are faster, it is also true that they are still further away from the original idea.

Why the evolution? Well, this is the way Ford does things. And now Ford, not Shelby, is producing the "Shelby" cars. Hardly had the ink dried on the 1968 press releases for the Shelby cars before Ford was offering several of the Shelby figerglass trim pieces in a trim package for the Mustang called GT/CS (California Special). Ford seems to have two principles of operation exemplified by all this: (1) if you've got a good thing, make it bigger and it'll be still better, and (2) if you hit upon a look that's popular, make all your cars look that way.

But Shelby still retains some control over the cars that bear his name, and to try to recapture some of the distinctiveness the name did imply—and to counter that high performance item from Chevrolet, the Camaro Z-28 (see p. 69)—has persuaded Ford to offer a new high-performance package for the GT 350. This package is the subject of this test, although we also tried a regular 350 complete with automatic transmission and air conditioning to see what the basic car is like these days.

The GT 350's standard engine is a hydraulic-lifter 302, developing 250 bhp @ 4800 rpm and featuring smooth, torquey performance with a smooth, slow idle. To the basic price of $4287 for the cooler version, the high performance package adds $692 and includes these items: special cylinder heads with larger valves (intake 1.875 in., exhaust 1.600 vs. 1.773 and 1.442), higher compression ratio (11.0 vs. 10.5), an aluminum high-riser intake manifold with a 715-cu-ft/min Holley 4-barrel carburetor (vs. an Autolite rated at about 650 cfm), a camshaft that still works with hydraulic valve lifters but gives greater duration (6° greater intake, 8° greater exhaust), limited-slip differential, stiffer rear springs and supplementary anti-windup leaves added to the front underside of the rear springs. No changes are made to the front suspension. Official figures weren't available at the time we tested it but the factory people estimated it as having 315 bhp @ 5000 rpm and 333 lb-ft torque @ 3800.

In operation the tweaked 302 is very little noisier than the mild one; with hydraulic lifters there's no valve clatter and the pleasant exhaust note (achieved with silencer nodes rather than the usual mufflers) is just a bit throatier. Instead of the smooth 600-rpm idle there's a rather lumpy 800-rpm speed but otherwise the engine is thoroughly tractable and without a sign of temperament. It has a useful rev range up to 6000 rpm and will touch 6400 before the valves float— in contrast to the standard engine's top limit of 5100. Hydraulic lifters have, indeed, come a long way.

Through the ¼-mile the hot GT 350 is a close match for the Z-28. It gets off the line in a much stronger manner and thus jumps well ahead of said Chevrolet; the Z-28 is just passing the 350 at the ¼-mi mark and going 6 mph faster— demonstrating that it is developing more power. The Ford is then limited by engine speed to a top of 119 mph (6400 rpm) whereas the Z-28 goes on to 132 mph (7100 rpm). Both the Z-28 and the hot 350 had 4.1:1 final drives.

Ford's own 4-speed manual transmission is standard in the 350. Its 10.5-in. clutch transmits the torque without slip or chatter and with reasonable pedal efforts, and it caused no problems in shifting at the peak engine speed. Shift linkage is precise but a little stiff and long of travel; synchromesh is unbeatable but slightly obstructive.

Power-assisted steering and brakes in the 350 follow expected U.S. practice: too much assistance, not enough feel. The 350 is a reasonably neutral, flat-cornering car but driving it fast over winding roads is a tricky proposition until one learns to steer and brake without the benefit of feedback information from the tires. Steering is reasonably quick at 20.3:1 overall, and brake performance isn't bad with the standard disc/drum setup; but all the fun is taken out by the dumb, numb power assists—both of which must be considered mandatory options because of the car's weight.

For the driver, the GT 350 (and any Mustang, of course)

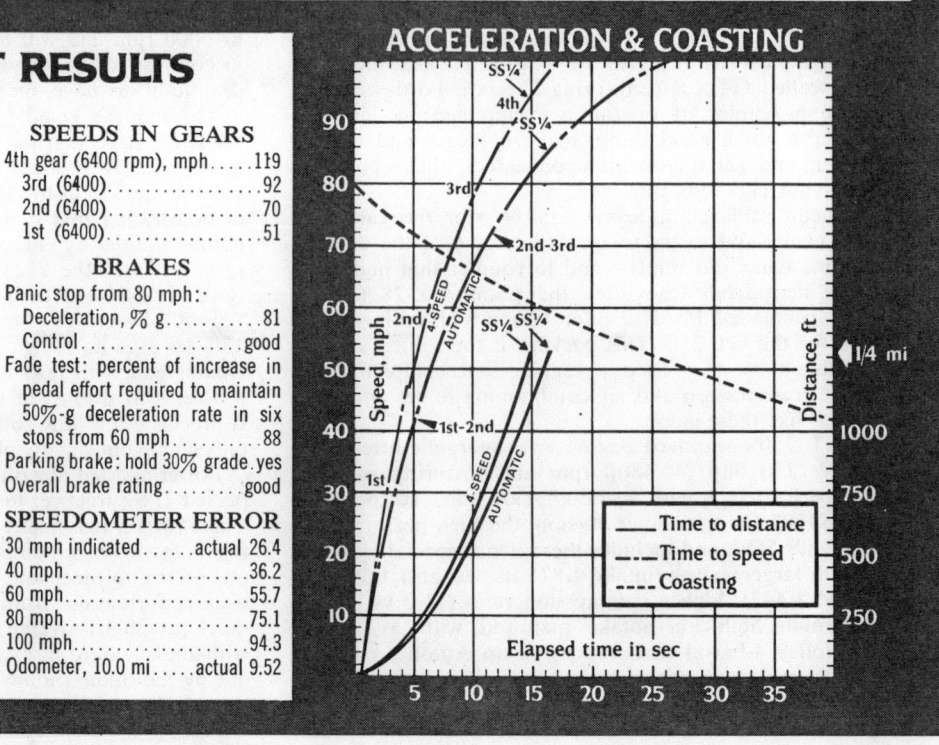

ROAD TEST
SHELBY GT 350

SCALE: 10" DIVISIONS

PRICE
Basic list..................$4287
As tested.................$5368

ENGINE
Type...................V-8, ohv
Bore x stroke, mm....101.6 x 76.3
  Equivalent in.......4.00 x 3.00
Displacement, cc/cu in....4949/302
Compression ratio..........11.0:1
Bhp @ rpm..............315 @ 5000
  Equivalent mph................93
Torque @ rpm, lb-ft....333 @ 3800
  Equivalent mph................78
Carburetion..........one Holley 4V
Type fuel required........premium

DRIVE TRAIN
Clutch diameter, in...........10.5
Gear ratios: 4th (1.00).....4.11:1
  3rd (1.29)................5.30:1
  2nd (1.69)................6.94:1
  1st (2.32)................9.53:1
Synchromesh...............on all 4
Final drive ratio...........4.11:1

CHASSIS & BODY
Body/frame: unit steel construction
Brake type: 11.3-in. vented disc front, 10 x 2.25-in. drum rear; power assisted
  Swept area, sq in...........325
Wheel...........steel disc, 15 x 6JJ
Tires...Goodyear Speedway E70-15
Steering type....recirculating ball
  Overall ratio..............20.3:1
  Turns, lock-to-lock...........3.7
  Turning circle, ft...........37.6
Front suspension: unequal-length A-arms, high coil springs, tube shocks, anti-roll bar
Rear suspension: live axle on multi-leaf springs, adjustable tube shocks

OPTIONAL EQUIPMENT
Included in "as tested" price: performance & handling package, power steering & brakes, AM radio, inertia-reel harnesses, folding rear seat, tilt steering wheel
Other: supercharger, alloy wheels

ACCOMMODATION
Seating capacity, persons.......4
Seat width,
  front/rear.......2 x 21.0/2x16.0
Head room, front/rear....40.0/35.0
Seat back adjustment, deg......0
Driver comfort rating (scale of 100):
  Driver 69 in. tall.............80
  Driver 72 in. tall.............80
  Driver 75 in. tall.............75

INSTRUMENTATION
Instruments: 140-mph speedometer, 8000-rpm tachometer, oil pressure, ammeter, fuel level, water temp, clock
Warning lights: high beam, directionals, alternator, brake fluid loss, emergency flasher

MAINTENANCE
Crankcase capacity, qt.........5.0
  Change interval, mi.........6000
Filter change interval, mi.....6000
Chassis lube interval, mi....36,000
Tire pressures, psi..........24/24

MISCELLANEOUS
Body styles available: coupe as tested; convertible
Warranty period, mo/mi: 60/50,000

GENERAL
Curb weight, lb..............3335
Test weight..................3640
Weight distribution (with driver), front/rear, %....55/45
Wheelbase, in.................108.0
Track, front/rear........58.1/58.1
Overall length...............186.8
  Width......................70.9
  Height.....................51.8
Frontal area, sq ft...........20.4
Ground clearance, in...........4.9
Overhang, front/rear....33.2/40.5
Usable trunk space, cu ft......5.6
Fuel tank capacity, gal.......16.0

CALCULATED DATA
Lb/hp (test wt)...............11.6
Mph/1000 rpm (4th gear).....18.4
Engine revs/mi (60 mph).....3260
Piston travel, ft/mi..........1630
Rpm @ 2500 ft/min...........5000
  Equivalent mph..............93
Cu ft/ton mi..................155
R&T wear index................53
Brake swept area sq in/ton....183

ROAD TEST RESULTS

ACCELERATION
Time to distance, sec:
  0–100 ft......................2.7
  0–250 ft......................5.5
  0–500 ft......................8.4
  0–750 ft.....................10.4
  0–1000 ft....................12.4
  0–1320 ft (¼ mi)............14.9
Speed at end of ¼ mi, mph....94
Time to speed, sec:
  0–30 mph.....................2.8
  0–40 mph.....................3.7
  0–50 mph.....................4.9
  0–60 mph.....................6.3
  0–70 mph.....................8.6
  0–80 mph....................10.6
  0–100 mph...................17.1
Passing exposure time, sec:
  To pass car going 50 mph....3.8

FUEL CONSUMPTION
Normal driving, mpg........11–15
Cruising range, mi........190–240

SPEEDS IN GEARS
4th gear (6400 rpm), mph.....119
3rd (6400)....................92
2nd (6400)....................70
1st (6400)....................51

BRAKES
Panic stop from 80 mph:
  Deceleration, % g...........81
  Control...................good
Fade test: percent of increase in pedal effort required to maintain 50%-g deceleration rate in six stops from 60 mph.........88
Parking brake: hold 30% grade..yes
Overall brake rating........good

SPEEDOMETER ERROR
30 mph indicated......actual 26.4
40 mph.......................36.2
60 mph.......................55.7
80 mph.......................75.1
100 mph......................94.3
Odometer, 10.0 mi...........9.52

ACCELERATION & COASTING

TWO SHELBY GT 350s

has a far less than perfect seating position. The seatback is too vertical, not adjustable (except by an owner modification of the stops) and the wheel is too close to the chest. On the other hand, the inertia-reel Y-shaped harnesses that come down from the excellent built-in rollbar are probably the best body restraint system now used in a production car. These are, incidentally, intended for use in conjunction with regular lap belts. All controls are easy to reach and would be even without the freedom of movement afforded by the inertia reels; but it isn't possible to roll down the right-side window without much stretching in this wide car. Vision to the front is impeded by the oppressively high hood and to the rear it's really poor because of large blind pillars. One nice item in the "visibility" category is the sequential brake lights (from the inside, symmetrically, zoop-zoop-zoop)—a safety item, we feel sure.

Instrumentation in front of the driver is readable if overstyled, but the supplementary oil-pressure gauge and ammeter installed in the central console are too far away, too low, not angled and poorly lighted at night.

Interior finish, slightly modified from standard Mustang materials and design, is pleasant and fairly restrained for an American car. The "+2" seat in the rear is pretty cramped but, when made folding by an extra $65, supplements the small trunk with extra luggage space. In short, however, this Mustang body is—like most American bodies—grossly large on the outside for the amount of passenger and luggage space it encloses.

OUR SECOND test car was a standard-engined, automatic-transmissioned, air-conditioned rent-a-car from the Hertz fleet that set us back $15/day and 15¢/mile. We found it to be almost pure Mustang in performance and handling—just a hair stiffer and faster than the 230-bhp 302 Mustang *Car Life's Ponycar* book reported on. Its 3-speed automatic transmission, though smooth and efficient, is not nearly as "manually controllable" as the ad writers would have us believe: "2" gets 2nd gear at most any speed, but "1" doesn't get 1st gear until the speed drops below 20 mph—and upshifts are accomplished only after considerable delay from the time the lever is moved. In this car the customer is getting the Shelby look combined with most of the luxury trim items, disc front brakes, big tires and styled wheel covers that the go-for-broke Mustang customer would order anyway, and it costs roughly $500 more than the comparably equipped Mustang.

PERHAPS THE FAIREST summary of these cars would be to say that IF they're the sort of cars to which you are attracted, they do the job well enough. We know from long experience with cars of this general type that it is well-nigh impossible to take a heavy, bulky car with an unsophisticated chassis and feel-less power assists and turn it into a truly sporting, responsive piece of machinery. In the early days of Shelby Mustangs the GT-350 seemed a serious and partially successful attempt to make a sporting car out of the Mustang; but now if anything it accentuates most of the Mustang's inherent shortcomings. And, what's worse, the Camaro Z-28 is a better example of the same sort of car—for less money.

We understand that Ford plans to put the mid-engine Mach II into production. Now *that* would be more like it.

Standard GT 350 engine has special valve covers and air cleaner but is essentially a stock Mustang 302 engine.

Optional engine features aluminum "high-rise" intake manifold, large Holley carburetor and more radical valve timing.

24 HEURES DU MANS 1969

Ford's 4th Straight

BY CYRIL POSTHUMUS

"THE TUMULT AND the shouting dies, the captains and the kings depart..." Le Mans is long over now, and you all know about Ford's fantastic fourth win and Porsche's desperately narrow defeat. But what a great race it was, what a turn-up for the book, what an incredible finish. There have now been 37 Le Mans 24-hr races, each and every one with its dramas, heartbreaks and tedium, but you'll have to go back a long time to find one matching Le Mans for sheer sensation and sustained interest.

It's hard to love Le Mans, all the same. Many pros in racing, and most journalists and photographers hate the place. It's smelly, dirty and overcrowded; the French officials, "men, vain men, dressed in a little brief authority," delight in being awkward; there are far, far too many interfering police around (on British circuits they control traffic, that's all), while the all-pervading air of crude commercialism taints the very real excitement of the battle between great marques like Porsche, Ford, Ferrari and Matra.

One wonders how paying spectators endure it in their noisy, dusty enclosures, with the dubious amenities of stinking 18th-century latrines and the added insult of having to stand in line *and* pay for them; the eternal grubby sand to flounder through; the hot, sickening wafts from the hamburger and hotdog stalls, and the hot, sickening prices the vendors have the nerve to charge for everything. There has to be some huge counter-attraction at Le Mans to fill the enclosures to bursting point with half a million spectators or so each year, and of course there is—the race itself, the savage pace, the harsh, thrilling music of multi-cylinder engines at work, going on and on through the day, the night, and through the day again to the climax and relief of the finish.

Of course it all began long before the traditional Saturday start on the traditional second weekend in June. This year the all-important national elections *(Pompidou ou Poher pour Président?)* mucked up the status quo and nearly shifted the whole race to another weekend, but eventually they settled for a 2 p.m. start instead of the usual 4 p.m. Then Porsche looked like wrecking the status quo even more thoroughly by withdrawing all their cars, and as they were fielding about 25 percent of the entry that mattered this would have been calamity from all angles. Smoldering over the FIA's cavalier ban on airfoils and "separate aerodynamic surfaces," the Germans demanded the right to use the suspension-controlled flaps on their big 4.5-liter flat-12 917s, declaring that the car was designed around them and demonstrating very forcibly before some splendidly frightened officialdom that they were unsafe without them.

Having clinched the Manufacturers Championship anyway, Porsche could play "san fairy ann" and tell the organizing AC de l'Ouest to go *étoffe* themselves, even though they badly wanted to win Le Mans and had seven very well prepared cars entered to try to do that very thing. In contrast, the ACO, already confronted with the withdrawal of four works Alfa Romeo 33s, three works Lolas, two Lancias and two Abarths, obviously wouldn't let such animators of the course as seven factory Porsches go without a struggle, even if it helped Matra's and Alpine's chances. A fine old *brouhaha* it was, with petitions and special meetings thrown in, ending in a climb-down compromise and permission for Porsche to race their 917s with the moveable flaps in operation, though the 908s had to wear theirs in a fixed position.

That settled, how did the field look? On paper, outright victory lay between Porsche, Ferrari, Matra and Ford, with inevitable qualifications on strength of entry, know-how, determination, resources, reliability and optimism. Porsche looked truly formidable. Four big 917s in all were listed, three works cars of which one served as reserve and non-started, and one private one, the first to be sold, for a big, cheery English lover of big hairy cars, amateur John Woolfe. And three 3-liter, flat-8 908s, two closed, one open, all of which put odds heavily on the first German victory at Le Mans since 1952, reinforced by staggering practice times in which, despite the Ford chicane just before the pits, Rolf Stommelen in a 917 lapped the 8.35-mile Sarthe circuit in 3 min 22.9 sec, which is 148.49 mph and faster than Denny Hulme's 3:23.6 record of 1967, set without the chicane in the 7-liter Mk 4 Ford. They say the 917 was doing almost 240 mph down the Mulsanne Straight. Some Group 4 car!

In face of such opposition, it was comforting to see Ferrari, the old hands at Le Mans, back again, although only two 312Ps, beautiful creations though they are, was cutting things a bit lean. Both had coupe tops, and one was all-new, though 420 bhp of 3-liter V-12 engine paled somewhat before the 585 bhp of Stommelen's Porsche. Matra put in four cars *pour la France,* all 3-liter prototypes with F1-based V-12 engines; they'd had a lot of trouble with crashes and injured drivers beforehand but were very determined about Le Mans. Three were open, one an all-new 650 for Jean-Pierre Beltoise/Piers Courage, two were older 630s for Johnny Servoz-Gavin/Herbert Muller and Nanni-Galli/Robin Widdows, and the fourth car was last year's 630 coupe for Jean Guichet/Nino Vaccarella—a fine international assortment of talent.

Two of the Fords were, of course, our old friends the Gulf-sponsored JW Automotive GT40s, with Jacky Ickx/Jackie Oliver in last year's winning car, and David Hobbs/Mike Hailwood in the second. There were also four privately-entered GT40s—one Alan Mann-prepared for owner Malcolm Guthrie and Frank Gardner, one for Peter Sadler/Paul Vestey, one a German entry for Kelleners/Jost, sponsored by the *Deutsche Auto Zeitung,* and one a French entry for the ASA-Esca organization. All Fords had full 5-liter engines—you can't give anything away in Group 4 these days.

With the Lola works withdrawal, the Daytona winners were represented only by the Scuderia Filipinetti's Traco Chevy-powered T70 GT Mk 3b for Jo Bonnier and that Kansas City character Masten Gregory—a formidably experienced pair. As it happened, the very same Ferrari 275-LM with which Masten won Le Mans with Jochen Rindt in 1965 was again running, again entered by NART (North American Racing Team, as if you didn't know) and driven by unemployed Alfa man Zeccoli and American Sam Posey. NART entered two other cars, a brand new Ferrari GTB4 Daytona and a 2-liter Dino, but alas and alack, Ricardo (no relation) Rodriguez in the Dino spun in front of Bob Grossman in the GTB4 on the Mulsanne brow, the pair inscribing their fate most legibly in black rubber as they bounced and rebounded off the new Armco barrier which has been erected along the entire straight. Neither Ferrari was deemed race-able after this drastic encounter, so Sam Posey transferred to the old 275LM, Bob Grossman just watched, and Le

The winning Ickx/Oliver Ford GT40 comes down the home stretch, finishing just this far ahead of the Herrmann/Larrousse Porsche.

The 12-cyl Porsche 917 led for hours and hours, set fastest lap, but faltered before the finish to hand the victory to Ford.

LE MANS 1969

Mans old hand Luigi Chinetti found himself with one car to team-manage instead of three.

Apart from the Lola, Filipinetti had two other charges—a real rumbleguts 7-liter Chevrolet Corvette Stingray coupe for Frenchman Henri Greder and Swiss up-and-comer Reine Wisell and a 275GTB Ferrari for Rey/Haldi. Then there were four Alpine-Renault 3-liter prototypes with rather gutless Gordini-developed V-8 engines. A twin-cam version, promised for months, was not ready so they had to rely on 310 meager horsepower, plus mild enterprises such as rear-mounted radiators and electronic stabilizers which restricted suspension travel at high speed. The drivers would have preferred 100 more hp.

Below the 3-liter mark there were Porsche 910s and 911s galore, two Belgian-entered Alfa Romeo 33s, hordes of blue Alpines in 1.5-, 1.3- and 1.0-liter sizes to guarantee at least one French victory, an Abarth, and the mighty British effort —one 2-liter Climax-engined Healey, one 2-liter BMW-engined Chevron, one 2-liter BRM-engined Nomad, a 1.3 Ford-engined Piper and a 1.3 Unipower. Neither the pretty Piper nor the homely Unipower, both unready and trouble-dogged, could qualify, which left the Healey, Chevron and Nomad, plucky efforts each, to represent the country whence came the winning Bentleys, Jaguars and Aston Martins of yesterday. Never mind, we've got Harold Wilson...

Practice, held in close, heavy weather, revealed several things, Porsches, their tendency to leak oil, the disastrous effects of getting 1st gear instead of 3rd, after which Woolfe's car needed an engine change; wheel and ventilation troubles *chez* Matra, engine worries *chez* Chevron; run bearings and withdrawal by the French-entered GT40; manifest unreadiness of the Piper—but what a looker; front spoiler troubles *chez* Ferrari, cured by throwing them away; bird trouble for Bonnier, sprayed with a largesse of feathers and gore when one jammed in a cockpit ventilator; a cracked bellhousing on the Hobbs/Hailwood GT40 which ruffled JW serenity; and the fact that superb ability on skis does not make an instant racing driver, as demonstrated by Jean-Claude Killy when he rolled his 1.5 Alpine at "Indianapolis" on his second lap . . .

So, AT LAST, The Day, sunless but hot, with air as thick as the gendarmes. Two hours before starting time the place was packed to bursting point with public and police, while over the public address system, in between much Gallic ado about nothing, they were playing that damnable song *"Les Vingt-Quatre Heures du Mans . . ."* Now came the finest hour for the gendarmes and petty officials with *carte blanche* to boss everyone around, to motion innocently and often prettily hanging legs back onto the pit counter, to check passes, and generally be unpleasant, while the public opposite whistled its customary derision.

Then engines started one by one, and soon the air vibrated with revving engines. The phalanx of silver-white Porsches at the front end of the long line of cars growled angrily in unison, the lovely red Ferraris joined in, while simultaneously by signal, the four blue Matras broke into a superbly bestial paeon of V-12 noises, causing all nearby to clap fingers hastily to ears. Rodriguez rolled up late and got a warm cheer, Frank Gardner chatted cheerily to Guthrie, Beltoise talked busily with his hands, Alpine mechanics ran hither and thither, and the *flics* gaped stupidly. Then came the order, *"Arrêtez moteurs"* and silence—apart from the excited chatter of the multitude—as the drivers took up their places on the road opposite for the running start so traditional to Le Mans yet soon, perhaps, to change because of safety harness problems.

Two o'clock, the *tricouleur*, and the *départ*. Forty-four drivers sprinted, and one ambled, across to their cars. The ambler was Jacky Ickx ("the last lap counts, not the first"), who was to win the race by bare yards yet who squandered that distance and much more by his leisurely getaway. His GT40, in similar mood, took its time to fire, joining in long after Stommelen had stormed away in a fine 45 degree slide to head a long string of Porsches on the first, perilous round. Past the pits he slammed, Elford's 917 at his tail, three 908s, the Lola, another Porsche and Hobbs' GT40 following and then, while an Alfa, two Matras and an Alpine also blatted through, back towards White House there came in a terrible flash, a tower of flame—and no more cars.

John Woolfe had lost his fierce Porsche 917 at White House; the car went off the road into a bank, bounced back upside down and exploded. Right behind, Chris Amon's Ferrari 312P charged into the blazing fuel tank; frantically pressing the Ferrari's extinguisher apparatus he baled out, getting away with a bad fright and slight burns. Other cars stopped at all angles behind the holocaust until firemen got it under control, and then trickled through, chastened at this first lap calamity. Poor Woolfe—whose co-driver, veteran Herbert Linge, had offered to take first stint—died almost at once.

Frank Gardner's GT40, sprayed in passing with blazing fuel, stopped at the pits; Jabouille's Alpine, too, was slightly damaged by flying debris; the rest went on with the race, Stommelen and Elford setting the pace in the big snarling 917s at around 139 mph. Before the first hour was up the Porsche armada held the first six places. Came the usual Le Mans phases; the first pit stops for driver-change, fuel, oil, water and maybe tires according to the cars' appetites; and the first retirements. A Porsche 910 and the pretty little Abarth spider were first to go, the boxy little Healey was dead unlucky, got a bolt through the radiator during the Woolfe accident, cooked its plugs and boiled away its water

Fourth went to France's big hope, the Beltoise/Courage Matra.

and its chances. Gardner's GT40 needed a radiator change and, more significantly, Stommelen's flying 917 developed an oil leak and a smoke screen, losing time and the lead in efforts to remedy same.

That gave the sleek open 908 of Siffert/Redman a turn at leading, though Elford/Attwood soon thrust their bigger 12 ahead, manfully keeping the revs down to 8000, and pegging it at a mere 218 or so mph down Mulsanne. But the Stommelen/Ahrens sister car was really in trouble; a gearbox oil seal had gone, and its smoke trail was getting thicker, so that the battery of 908s in reserve must have comforted ex- and present team managers Huschke von Hanstein and Rico Steinemann, both Teutonically deadpan in the pits. Yet it was the Siffert/Redman 908, one of their big hopes, which went next, contracting a seized gearbox just before the fourth hour. *Zut alors* . . . Behind, the Gulf Fords lay 7th and 8th, traveling nice and steadily, straining nothing.

Elsewhere attrition was making hay. The blaring red Ferrari 312P of Rodriguez/Piper, depressingly way back after the lap 1 fracas, lost more time investigating gearbox trouble, though the engine sounded as if it would go on forever. The mechanics on the Gardner/Guthrie GT40, having a harder time than the drivers, now changed a driveshaft; Grandsire's 3-liter Alpine overheated, then blew its top, leaving room at the pits for other ailing Alpines. Slotemaker's 2.5 Alfa 33 broke its tail, then an oilpipe, and went out; Lanfranchi's Nomad-BRM went out with oil in the wrong places; Wicky's Porsche 911 threw a rod; the Filipinetti Ferrari 275GTB took an oil too early and was disqualified, and Gosselin's 2-liter Alfa 33 had braking trouble at the Ford chicane, went straight through it, and was wrecked. The 45-car starting list was beginning to look sick.

By the sixth hour, quarter distance, the leaders were Elford/Attwood (Porsche 917), averaging 136.1 mph, two laps ahead of Mitter/Schutz (Porsche 908), and third, *mes amis,* lay the Beltoise/Courage Matra; fourth were Lins/Kauhsen (Porsche 908) and fifth Nanni-Galli/Widdows (Matra), with the Gulf GT40s of Hobbs/Hailwood and Ickx/Oliver dogging their tracks. The other 908s were having their troubles, Herrmann/Larrousse losing 31 minutes changing a front hub, while Mitter rolled a tire off at Mulsanne—it all costs time and helps the opposition—as instanced when Beltoise enjoyed second place for a short time, only to drop to fourth with a long pitstop.

So the pattern of the race developed. Porsches out in front, nursing as much as they dared; the slower Matras trying hard; the GT40s just sitting, for this was Le Mans and did not the Book of Wyer counsel wait-and-see, even when the Sage himself was absent by his sick wife's bedside? The attrition continued, tidying up the lap charts of those who had to try and keep one. Gardner/Guthrie packed it in. So did the yellow Chevron of Brown/Enever (gearbox); the Bonnier/Gregory Lola needed new gaskets; Widdows' Matra dropped from the first six with fuel injection trouble; the Sadley/Vestey GT40 departed with electrical maladies, and the Stommelen/Ahrens Porsche was an ominously long time at the pits, its fine pace broken. Someone must have smashed that *"Vingt quatre heures du Mans"* record, bless him, for we hadn't heard it since the start. By now the lights were on and it was Le Mans *allumé*. The gaily-lit restaurants, *le village,* the fair, and the woods beyond offered varied diversions to the continuous boom of the cars; the race was young and there's still tomorrow, so hell . . . Yr hmble svt., I regret to record, simply went to bed.

One of the interesting things about Le Mans is to return on Sunday morning, fresh and defiantly shamefaced before the stalwarts who stayed all night, and wonder at the remarkable changes which the long run through the dark and the cold, misty morning can wreak on the race. The 24 cars left at midnight had been reduced by several important runners—the Stommelen/Ahrens Porsche had gone (clutch, etc.); the Servoz-Gavin Matra had gone (steering), the Greder/Wisell Corvette had gone (lack of gears); Masten Gregory in the rebuilt Lola had a rod go at full chat along Mulsanne; Alpine, having a terrible Le Mans, had lost all their V-8s one by one and also two 1500s so far; Udo Schutz touched the rail in the Mulsanne kink and practically wrote off his 908 but not himself (Hobbs punctured on the debris); and the Ferrari, alas, had gone too (transmission, etc.) after consuming pailfuls of oil.

But the surviving Porsche 917 of Elford/Attwood was still bombing around in front, averaging 132.123 mph by 10 a.m. Sunday and four laps up on the Lins/Kauhsen 908. Porsche still 1-2, but the Fords were there behind them, loping like patient wolves waiting for the travelers to tire. Ickx/Oliver lay third, 4 laps back, Herrmann/Larrousse (908) were fourth, Hailwood/Hobbs (GT40) fifth and Beltoise/Courage (Matra) sixth, each separated by a lap. Then Fate resumed its cruel game with Porsche. First it picked on the leading 917, bringing Attwood to the pits just after 10 with severe clutch slip. They doused it with gasoline and off he went again, hesitantly and trailing a dismal stream of smoke next time around.

The crowd buzzed, sleepy newshawks in the press stand perked up, and Gulf-JW team manager David Yorke's eyes

Pit action for the 3rd-place finisher, the Ford driven by Hobbs and Hailwood.

69

LE MANS 1969

gleamed. Small wonder, for a lap or two later Rudi Lins in the second place 908 stopped and handed over to Willi Kauhsen—who returned after a mere lap for urgent work on *his* clutch, then resumed racing very gingerly. He never reappeared, for the transmission seized at Mulsanne and one more Porsche had gone. Now Ickx/Oliver were second and gaining swiftly; by 11 o'clock they were leading, just as the Elford/Attwood 917 pulled in crestfallen to its pits. *Kaput* was the bitter verdict, and the once proud Porsche, now grimed, oily and broken, was pushed slowly and sadly away. The crowd clapped, cheered and whistled, and one would like to think they did it in sympathy for a magnificent effort, cruelly ended.

What next? Ten minutes later Ickx brought the Ford in for water and brake pads, and while halted Hans Herrmann in the 908 whipped past to retrieve the lead for Porsche. By the time Ickx was back racing, the Porsche was 47 sec ahead until it, too, stopped for fuel. Frenchman Larrousse took over and began a grim duel with the Ford, holding it off by bare yards until 12:30 and the last pit stops. Ickx shot in for fuel and out again, co-driver Oliver having to sit out the final drama, while Larrousse led until his pit stop two laps later. Then the ever youthful veteran Hans Herrmann set off, the full load of Porsche responsibility on his shoulders, just as Ickx tore through. The crowd could scarcely contain themselves at the excitement.

Next lap the Porsche and the Ford were nose to tail; 15 yards apart the lap after; shuttling back and forth as the last laps unreeled. With under an hour left, Herrmann came through ahead; next round Ickx repassed; the round after that Herrmann led again; then it was Ickx again as the warring pair came up on Hailwood in the 3rd-place GT40. Mike moved up beind his teammate and kept off Herrmann and then, in the middle of this desperate last-hour Grand Prix, a long, thin blue line of police moved in on the pit area and the grandstand crowd broke into derisive whistles and catcalls. Then Hailwood dashed suddenly to the pits—he was running out of fuel and was nearly rammed at 180 mph or so by an agitated Herrmann—bringing the Porsche back on to Ickx's tail.

A mere 15 sec before 2 p.m. and the finish, the pair fled past the pits, thereby having to cover another desperately tense lap while the flag, going out at 2 p.m. precisely, ended the race. At last they came into sight, blue 100 yards ahead of white, and it was Ford the victors from Porsche—just! After all that terrific tension it was light relief to watch the police form walls each side of the presentation dais, unattainable by Ickx and the winning car on one side and by co-driver Oliver and the jubilant Gulf-JW pit crew on the other. Oliver solved it by charging straight through the legs of the police, and the two triumphant Jackies met on the roof of their gallant GT40.

The results herewith tell the rest of the tale. Of how the ancient GT40s took 1st, 3rd and 6th places *and* won the Index of Thermal Efficiency, really gilding their place in the Hall of Fame; of how Matra covered themselves in glory in their second Le Mans with three cars out of four finishing. Of how a Ferrari *did* finish, the grand old 275LM ably driven by Zeccoli and Sam Posey, the first and only U.S. driver home. Of how Alpine saved the shreds of self-respect by winning their umpteenth Index of Performance with their last surviving car ... But whew, what a race!

24 HEURES DU MANS 1969

Le Mans, France—June 14-15, 1969

Driver/Driver	Car	Gr.	Miles
1 Ickx/Oliver	5.0 Ford GT40	4	3104.35
2 Herrmann/Larrousse	3.0 Porsche 908	6	3104.28
3 Hobbs/Hailwood	5.0 Ford GT40	4	3077.04
4 Beltoise/Courage	3.0 Matra 650	6	3071.76
5 Guichet/Vaccarella	3.0 Matra 630	6	3002.69
6 Kelleners/Jost	5.0 Ford GT40	4	2852.53
7 Galli/Widdows	3.0 Matra 630	6	2760.24
8 Zeccoli/Posey	3.3 Ferrari 275LM	4	2747.86
9 Poirot/Maublanc	2.0 Porsche 910	6	2604.54
10 Gaban/Deprez	2.0 Porsche 911S	3	2560.47
11 Lena/Chasseuil	2.0 Porsche 911T	3	2518.27
12 Serpaggi/Ethuin	1.0 Alpine	6	2443.21
13 Laurent/Marche	2.0 Porsche 911T	3	2401.10
14 Farjon/Dechaumel	2.0 Porsche 911S	3	2389.57

Distance: 372 laps of 8.35-mi circuit, 3104.35 mi. (Record: 3249.6 mi, Dan Gurney/A.J. Foyt, 7.0 Ford Mk 4, 1967.)

Avg speed: 129.39 mph. (Record: 135.483 mph, Gurney/Foyt, 1967.)

Fastest lap: 3:27.2, 145.419 mph, Elford, 4.5 Porsche 917. (Record: 3:23.6, 147.89 mph, Denis Hulme, 7.0 Ford Mk 4, 1967.)

Index of Thermal Efficiency: 1 Ickx/Oliver, 2 Hobbs/Hailwood, 3 Kelleners/Jost, 4 Serpaggi/Ethuin, 5 Beltoise/Courage.

Index of Performance: 1 Serpaggi/Ethuin, 2 Herrmann/Larrousse, 3 Beltoise/Courage, 4 Guichet/Vaccarella, 5 Ickx/Oliver.

Last Year's Winners: Pedro Rodriguez/Lucien Bianchi, Ford GT40, 2764.2 mi, 114.9 mph.

FORD GT MARK II

CONTINUED FROM PAGE 49

remain 11.6 in front and rear, and Girling BR single calipers are used at both ends of the car.

BODY: Only the midsection of the Mk I body remains the same in this year's Mk II. Front and rear sections are completely new and it is a tribute to their designers that absolutely no add-on tabs, fins or the like were needed this year. Only the adjustable spoiler at the rear was needed, and this was set at heights from 1.5 to 3 in. to suit different drivers.

The most striking body change is the front end. It is 9 in. shorter than last year's nose and is actually a "production" GT-40 nose with slightly higher fender humps for more wheel travel. Wind tunnel work, and Phil Remington's engineering intuition, said that the nose had to be as low as possible. And the shorter length gets the center of pressure back toward where it should be. The front end, therefore, is responsible for the lack of fins on the rear. It also saves 19 lb.

The rear end is higher, for clearance of the new vertical suitcase bins, by about 2 in. at its extremity. The wind tunnel was used to develop new ducting for improved engine, transaxle and brake cooling, and the new duct disposition is thus:

- Side scoops—low, right side: transaxle
  - —low, left side: engine oil
  - —high, both sides, upper 2/3: carburetor
    - lower 1/3: brakes
- Top deck scoops—added at Sebring: brakes

Louvers in the center of the rear deck direct air over the exhaust pipes and get that hot air out. The new tail weighs 37.5 lb vs 80 for the old one.

Other body improvements have been for convenience, such as a new access panel in the nose for getting at the spare tire and oil tank. Before, the whole nose had to be lifted.

Mark II Driving Impressions, by Ken Miles

"WHEN PROPERLY set up," Ken Miles says, "this is the easiest car in the world to drive. If not, it's awful—but then, this is typical of any really modern racing car. Small changes in chassis tune produce large changes in handling. The suspension is designed for a particular ride attitude; as speed goes up this attitude changes. We spent two days getting the car to handle right at Le Mans, found that the spoiler setting needed to be 2 in., not 1. Most critical thing is precise control of rear suspension geometry with 4-link arrangement used today—tires are still an area of much ignorance—even after arriving at basic suspension geometry with help of IBM, I have to get it adjusted on the track.

"It's a cooking engine. I can lug it down to 1000 rpm in 4th. When does it come 'on the cam'?—oh, about 3000 rpm! We babied them. The thing's safe for 7400 rpm, but we never exceeded 6200 in the race.

"It's a bloody oven inside! It takes a fair amount of physical effort to drive—steering is heavy—you brace yourself in banked turns so you can hold onto the controls. This is due largely to the fact that the car has been developed so far past its original state. The steering feeds back quite a lot, and I get big blisters...

"Cornering is pretty neutral, takes severe provocation to hang the tail out, and then it's only briefly. She really wants to stay put. I say neutral, but that's my car. Ron Bucknum likes a little noseplow, Lloyd Ruby wants the tail hanging out—we get three different patterns of tire wear.

"The gearbox is easy to shift after broken in and has completely unbeatable synchro. It could be lighter if it weren't built around Galaxie internals, has an extra shaft to bring drive back through.

"Brakes are high-effort. Can't possibly lock wheels. They are our Achilles heel; there's just not room for a brake big enough. Running 1650° F, there's too much variation in temperature—I planned my driving so only one disc change would be necessary. Unfortunately, one of the new discs was bad and I had to stop again. At the end of the race, this set was in good condition, ready for another go.

"The seating is very comfortable, yes. Hard to get into, but no aches at all after the race. For rear vision, pick the mirror glass off the floor and hold it in your hand. The wipers work well at high speeds, but the washer hardly works at all.

"Throttle linkage is very important. It must achieve two things: it must be smooth, and progressive—slow at first opening, getting faster. I have these qualities, but I don't like the suspended pedal. Not natural.

Riding with Ken, I found that indeed the Mk II is a flexible car. It isn't quiet, of course, but with the big noises going out the back it's not horrendous; the ride is about that of a street Cobra. There's a realization of great structural strength, but there are rattles all over. And it IS possible to lug from 1000 in 4th. Acceleration in the indirect gears, once past 3000 rpm, is simply indescribable.

Summary

THE FORD GT Mk II, though its beginning was in England, has evolved into a distinctly American car. Contrasted to its contemporaries, it compares with them just as an American sports car does with European ones—the big point of difference being that the American car has a relatively heavy, large-displacement and slow-revving engine with unsophisticated valve gear. For the first time it proved that this power concept can be fully competitive with the light, high-output European engines in a long distance race like Le Mans.

If Ford is to continue racing, and it appears fairly likely that the company will, it is the new "J" car that will be campaigned as the prototype car. It is the logical evolutionary step from the Mk II, but having been drawn up on a clean piece of tracing paper, with the 427 engine designed in, it embodies all the Ford people have learned in their three years' experience with the GTs.

Miles and Ruby drove an open car at Sebring that was a prototype for the J car: it had a single-sheet aluminum underbody, which wasn't quite satisfactory but which pointed the way toward using the honeycomb aluminum structure the J car now uses. As a result of the use of this material in bulkheads (as well as other new structural efficiencies) the weight of the J car's underbody is only 169 lb, versus 360 for the Mk II's single-sheet unitized steel underbody.

Back at Ford Dearborn, out there at Shelby American, over there at Alan Mann and down there at Holman-Moody, everybody has crossed fingers. At this writing it was budget time in Dearborn. We all hope the budget says "go".

COBRA WINS

CONTINUED FROM PAGE 35

coupe driven by Dan Gurney and Bob Bondurant. This was the highest placing by a Grand Touring car at Le Mans and an achievement that had sometimes seemed impossibly far away.

The critical event in the 1964 international championship competition was the Tour de France rally which, because of another of the FIA's funny, funny regulations, counted as much toward the championship as the Sebring 12-hr race. Four Cobras were entered in the rally, two coupes and two roadsters, but all four had mechanical troubles of one variety or another and Ferrari clinched its fourth straight manufacturers' title. The final points totaled up to 84.6 for Ferrari and 78.3 for Cobra.

Also in 1964, Shelby-American gave its full support to the USRRC series and again won the manufacturers' division in the over-2-liter class.

FOR 1965, the handwriting was clearly on the wall for Ferrari. Everything seemed set against the Maranello firm from the beginning, even the rules. In past years, Ferrari had things pretty much his own way with FIA and obtained homologation of models that met the regulations only through the most magical of loopholes. This year Ferrari couldn't obtain approval of the models he wanted to have classified as GT machines and very possibly outsmarted himself, as this left the firm without an appropriate weapon with which to do battle with the Cobras. The Ferrari GTO was out of production, the 250-LM wasn't homologated and the GTB in production form simply didn't have the beans to get the job done.

So the 1965 season opened with the Cobras in an almost unassailable position. The season opened at Daytona again and a coupe driven by Jo Schlesser and Hal Keck headed a 1-2-3 Cobra sweep. Two years earlier, a pair of roadsters had suffered humiliating mechanical failures; last year the coupe had led only to fail two-thirds of the way home. So the victory this year was extra sweet. And this was to establish the pattern for Cobra's season.

At the Sebring 12-hr race, Schlesser was teamed with Bob Bondurant and again Cobras swept to an almost unopposed victory. In the next race in the championship series, the coupes carried the battle to Italy, where Bondurant teamed with another young American driver, Allan Grant, to win the 1000-km race at Monza. Sir John Whitmore followed that up with a first in the Tourist Trophy at Oulton Park.

No Cobras were entered at the Targa Florio, but two weeks later Bondurant finished second in the GT class at Spa, Belgium, being led across the line by a Ferrari GTO driven by Sutcliffe after suffering from an ignition fault that kept the Cobra on seven cylinders through most of the race. Bondurant then teamed with Jochen Neerpasch to win the GT class at the Nurburg Ring 1000-km race, one of the team's best victories during 1965, and followed that up with a victory at the Rossfeld Hillclimb. Then came Le Mans. If the Cobras had been lucky anywhere else, they lost all touch with Dame Fortune in the 24-hr race. Of four coupes to start the race only one was still running at the finish. The battered and sick car driven by Jack Sears and Dick Thompson limped into 11th overall, led to the finish by a Ferrari GTB that had been nowhere in sight when the full strength of the Cobras was in evidence.

Two weeks after Le Mans, Cobras clinched the championship by winning the GT class at the Reims, France, 12-hr race. Unopposed by Ferrari GT cars, a coupe driven by Bob Bondurant and Jo Schlesser finished fifth overall behind four Ferrari prototypes.

About the victory, Carroll Shelby said, "I don't believe anything in my own racing career, even winning at Le Mans, has been more gratifying than winning this championship."

THERE HAVE been a total of six of the Cobra Daytona coupes built, and all six are still in existence though they have been campaigned about as hard as it is possible to campaign a racing car. They have proved themselves to be both fast and durable and were demonstrably superior to the GT cars that Ferrari has fielded the past two years.

The coupes are slightly larger in bulk than the roadster version of the Cobra but weigh almost exactly the same, about 2400 lb. They are built on the same chassis as the roadsters, as required by the regulations, but the chassis is slightly stiffer in the coupes through the addition of stiffening members (permitted by the rules) and the nature of the coupe body. As a result of the more rigid chassis, the somewhat dated suspension (transverse leaf springs do seem a bit naive these days) was able to more nearly perform the function it was designed to do and the coupe consequently handled somewhat better than the more flexible roadster. More important, perhaps, was the improved aerodynamics of the coupe, which enabled it to attain top speed of about 170 mph compared with the flat-out maximum of 155 for the roadster.

The engine and running gear of the championship-winning coupe is the same as on the competition roadsters. As raced this year, the 289-cu-in. Ford engine was developing about 385 bhp. This engine has been specially breathed upon at the Cobra works and uses four double-throat Weber carburetors, a modified head that includes bigger valves for better breathing, a revised combustion chamber shape and higher compression. The tolerances of the engine are also opened up throughout, the oil capacity is larger and the oil pressure increased by setting the pump for greater output.

The interior of the coupe is stark, there is no sound deadening or padding, the seats are comfortably contoured but sparingly padded and the noise level is high.

The future of the championship-winning coupes is not clear. The FIA's rules for 1966 have not yet appeared in their final version, but it seems likely that the coupes would be eligible to participate in the new Category A, Group 4 because they were previously homologated in the comparable class. However, the Ford GT-40 is being aimed at that same class for 1966 and the Daytona coupes would hardly be competitive in that company. Nor is it known just now how they would fit into SCCA racing in this country. They could be run in the manufacturers' championship section of the U.S. Road Racing Championship, but on the typical U.S. road racing circuit, the 1966 Cobra II with a 427-cu-in. engine would probably be quicker.

The coupes are for sale, however, and though the price hasn't been fixed, it is expected to be about $10,000 the copy—or perhaps a bit less if you don't mind the somewhat ratty as-raced condition. That isn't a bad price, considering the competition roadsters costs almost that much, and you would be getting a car of rare distinction.

You might not win any races with it, but, man, wouldn't it turn them on down at the drive-in?

its single 4-barrel with my left hand, operate the solenoid switch with my right and immediately juggle the linkage. A chokeless Grand Prix Bugatti I had years ago was more of a problem because it had no starter motor and had to be started on the crank. The solution was to stuff a handkerchief into its Solex, work on the crank and jerk the handkerchief out (usually singed) as soon as it fired.

The driving position of the Cobra is comfortable, with plenty of room for tall people. The only criticism is that because of the size and location of the engine and transmission, the pedals are offset appreciably to the left and the shift lever is comparatively far back. However, these faults are easy to get used to and don't affect one's driving.

Obviously the location of a 7-liter iron engine in a 2500-lb car is critical, because one could easily have a nasty polar moment of inertia if it weren't positioned just where it should be. The front/rear weight distribution is 48/52% without driver. Considering the torque of the engine, one might expect the 11.5-in. clutch to be a real beast, but actually it is very light and it takes up very smoothly.

75

The transmission is one of the best I have ever come across in years of sports-car driving and one can find a ratio for any occasion among the four speeds. The engine idles at a steady 800 rpm when warm and the rumble shakes my garage, which is admittedly a bit Tobacco Road. Driving around the neighborhood, and not wanting to foul the nest so to speak, I just run it up to about 1500–2000 between shifts, which doesn't appreciably increase the decibel level.

A fortunate characteristic of the 427 engine is that, despite all the carburetion and the humper-humper-humper sound at idle, it is extraordinarily flexible and could probably pull a freight train at 2000 rpm. Not only does this characteristic make the car easy to drive, but it also means that you can keep the secondary barrels—which are the noise makers—out of contention. The steering is quick, at 2.5 turns lock-to-lock, and heavy when moving slowly because of the amount of rubber on the ground.

From the styling point of view, the car is exactly in keeping with its character because it looks mean as hell, whether coming or going. None of your sleek, sexy Italian styling for Carroll Shelby and his men, but just a bundle of brute force covered in aluminum wrapping shaped to fit, flared and bulged where necessary.

As far as creature comforts are concerned, they are minimal. The cockpit ventilation is inadequate, but then 400 plus horsepower represents a lot of heat to be disposed of. The noise becomes tiresome on a long trip, but there is nothing you can do about it because there is no room for adequate mufflers and the tail pipes, for the same reason, must exhaust in front of the rear wheels. The top offers adequate protection, but assembling it is a sort of Barnum and Bailey circus-tent type of operation. There is a heater, other than the engine, but it doesn't really seem to be necessary; the radio, which was installed at some time or another, has been removed by a previous owner—presumably because it couldn't compete with the exhaust noise.

Service accessibility is surprisingly good because the car is just basic machinery and therefore uncluttered by air conditioning, power steering and all the other devices that get in the way of a mechanic. However, one slight failing is that apparently you have to remove the right cylinder head to replace the battery.

Once you are away from human habitation, you are ready for the moment of truth, but it is essential to proceed with caution and respect. Unlike most other fast cars, it's not the 425 bhp but the 480 lb-ft of torque at 3700 rpm that gets you into trouble. It's torque that spins wheels, and the torque comes in very early indeed and without much warning at all.

At the time the 427 Cobra was introduced it was fashionable to record times from zero to a hundred and back down to a complete stop for fast cars. The record was then held by Aston Martin at slightly under 25 seconds, but the late and much lamented Ken Miles, who was a Cobra driver and developer, went out and did the job in 13.8 sec. Not only does the car go, but it also stops as well, and using today's rubber Ken certainly could have improved on his performance.

What happens is that you drive around slowly to get everything warmed up and then you put your foot in it in 2nd gear. Instantly it's transformed from a lumbering, heavy monster into something that could have escaped from Cape Kennedy. There is a colossal Blaaaaaaaaaaaaaa of noise, the tires smoke, the tach hits 6000, you are ready to do the same again in 3rd, and for an encore you can do it again in 4th a few seconds later if the car is still cooperating. The great thing about torque is that it doesn't really matter what speed you are going or what gear you are in—there's always a bit of urge left.

The purists tell me that all this sort of thing is rather crude, and I don't blame them for being jealous. I have never played automobile oneupmanship, but when people challenge me, I find that *427 Cobra* is a very heavy name to drop and the subject is changed instantly. Actually, the car is not crude at all by 1965 standards. Admittedly it is tricky to drive hard because of the enormous power, but the suspension is excellent on good road surfaces, the brakes are superb and the steering extremely accurate. On poor surfaces the car tends to leap about and scratch for traction and you can break traction very easily indeed when accelerating hard on any surface.

There is nothing particularly treacherous about the car's basic handling qualities, because it is strictly a neutral machine as far as oversteer and understeer are concerned. However, the whole question is somewhat academic because if you are feathering it through an 80-mph bend and start sticking your foot in it, you know what happens when 480 lb-ft or a multiple thereof hits the rear wheels, and it doesn't really matter what gear you are in. With two 4-barrels it's all too easy to activate a few more barrels than you actually need. When I wrote my impression of the Allard I described it as perhaps the most dangerous car I have ever driven, but I think the Cobra surpasses it. The trouble with the Allard lay in its split front axle, which could virtually give you oversteer and understeer in the same breath because the transition was instantaneous. The trouble with the Cobra is simply its power, which is no fault at all—it's just that there is a temptation to use it.

Some people have compared driving the 427 Cobra to driving one of the 1937 Mercedes-Benz W125 race cars, which in their ultimate state of development put out 646 bhp at 5800 rpm from a straight-eight, supercharged engine of 5660-cc capacity. Obviously this is not a fair comparison for many reasons, but it is an interesting one all the same.

To make the comparison fair, the Mercedes would have to be equipped with 1974 rubber. With a curb weight almost exactly the same as the Cobra, the Mercedes would undoubtedly outrun the Cobra on long straights because W125s were timed at 193 mph at Spa and could have gone up over 200 mph given a suitable axle ratio. The high maximum speed comes from the 646 bhp at 5800 rpm, and it would tell over the Cobra's 425 bhp. On the other hand, the Mercedes would be strictly limited on the average road course by its brakes and suspension, and the victory would go to the Cobra.

Buying oddball cars is a risky business, because you never really know whether you are going to like the machine until you have lived with it for a while. After six months of living with a 427 AC Cobra, I seem to like it more every day, and there is great satisfaction in knowing that there is absolutely nothing on the roads (except another Cobra) that could possibly catch it, at least up to 150 mph.

R&T ROAD TEST
COBRA 427

PRICE
List price, west coast, 1966....$7495

MANUFACTURER
AC Cars Ltd
Thames Ditton, Surrey, England
Shelby American Inc
Los Angeles, Calif.

GENERAL
Curb weight, lb	2530
Test weight	2890
Weight distribution (with driver), front/rear, %	est 49/51
Wheelbase, in.	90.0
Track, front/rear	56.0/56.0
Length	156.0
Width	68.0
Height	49.0
Ground clearance	4.4
Usable trunk space, cu ft	6.8
Fuel capacity, U.S. gal.	18.0

ENGINE
Type	ohv V-8
Bore x stroke, mm	108.0 x 96.2
Equivalent in.	4.24 x 3.79
Displacement, cc/cu in	6998/427
Compression ratio	11.5:1
Bhp @ rpm, gross	425 @ 6000
Equivalent mph	150
Torque @ rpm, lb-ft	480 @ 3700
Equivalent mph	92
Carburetion	two Holley (4V)
Fuel requirement	premium

DRIVETRAIN
Transmission	4-sp manual
Gear ratios: 4th (1.00)	3.54:1
3rd (1.29)	4.57:1
2nd (1.69)	5.98:1
1st (2.32)	8.21:1
Final drive ratio	3.54:1

CHASSIS & BODY
Layout	front engine/rear drive
Body/frame	steel tubular frame, separate aluminum body
Brake system	disc; 11.63-in. front, 10.75-in. rear; vacuum assisted
Swept area, sq in.	550
Wheels	cast magnesium, 15 x 7½
Tires	Firestone Wide Oval G70-15
Steering type	rack & pinion
Turns, lock-to-lock	2.8
Turning circle, ft	36.0

Front suspension: unequal-length A-arms, coil springs, tube shocks
Rear suspension: unequal-length A-arms, coil springs, tube shocks

INSTRUMENTATION
Instruments: 180-mph speedometer, 8000-rpm tachometer, 99,999 odometer, 999.9 trip odo, oil pressure, coolant temperature, ammeter
Warning lights: high beam, directionals

ACCOMMODATION
Seating capacity, persons	2
Seat width	2 x 20.0
Head room	33.0
Seat back adjustment, deg	none

MAINTENANCE
Service intervals, mi:
Oil change	2000
Filter change	2000
Chassis lube	4000
Warranty, mo/mi	3/4000

CALCULATED DATA
Lb/bhp (test weight)	6.0
Mph/1000 rpm (4th gear)	25.0
Engine revs/mi (60 mph)	2400
Piston travel, ft/mi	1515
R&T steering index	1.01
Brake swept area, sq in./ton	381

ROAD TEST RESULTS

ACCELERATION
Time to distance, sec:
0-1320 ft (¼ mi)	13.8
Speed at end of ¼ mi, mph	106.0

Time to speed, sec:
0-30 mph	2.4
0-40 mph	3.3
0-50 mph	4.3
0-60 mph	5.3
0-70 mph	7.4
0-80 mph	9.4
0-100 mph	13.0

SPEEDS IN GEARS
4th gear (6500)	162
3rd (6500)	126
2nd (6500)	113
1st (6500)	77

FUEL ECONOMY
Normal driving, mpg	12.0
Cruising range, mi (1-gal. res)	204

BRAKES
Minimum stopping distances, ft:	
From 60 mph	130
Control in panic stop	very good
Fade: percent increase in pedal effort to maintain 0.5g deceleration in 6 stops from 60 mph	nil
Parking: hold 30% grade?	no
Overall brake rating	very good

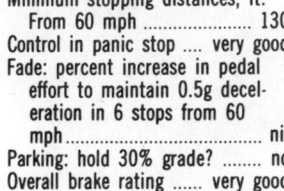

SPEEDOMETER ERROR
30 mph indicated is actually	30.0
50 mph	47.0
60 mph	57.0
70 mph	66.0

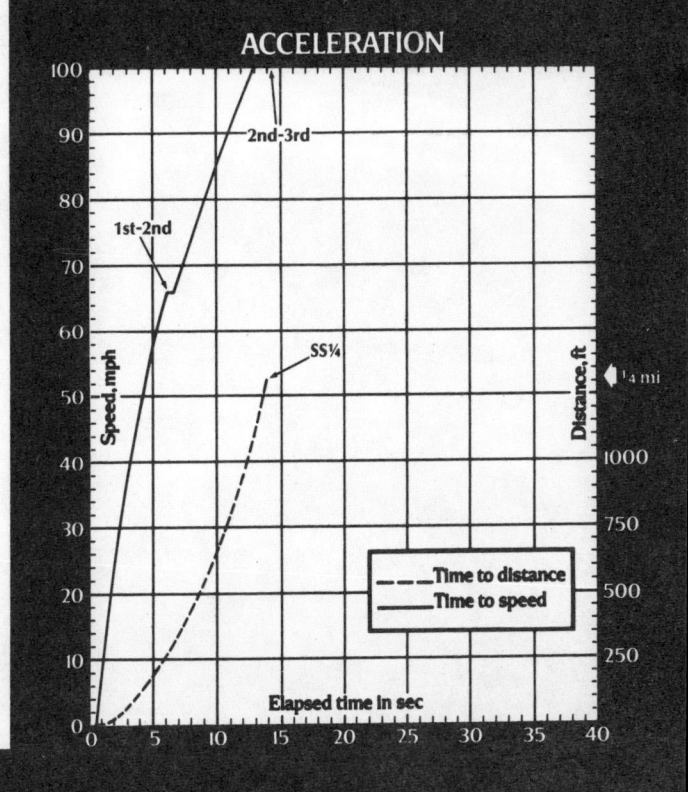

M IKE SHOEN OF PHOENIX, Arizona is the ultimate Cobra enthusiast. Not only does he have some half-dozen of these cars, but among them is the original FIA 289 Cobra team car raced by Phil Hill at the Targa Florio in 1964. Mike was kind enough to produce this car at Laguna Seca Raceway for us and so we reciprocated by producing Phil Hill and also Dan Gurney, who drove the sister car, so that the two drivers could reacquaint themselves with this famous machine.

Mike's enthusiasm is such that when he says he has Phil's original car, what he really means is that it is absolutely original down to the last nut and bolt. He even has the original tires, although the rears are too worn to use other than for display purposes. Of course, the car is in as-raced condition, complete

Salon

1964 FIA Cobra Roadster

Unsophisticated, but tough and very fast

BY TONY HOGG
PHOTOS BY JOHN LAMM

with dents and dings, and the firing order is still chalked on the underside of the hood where some mechanic wrote it many years ago. At first sight, we were somewhat disappointed by the car's condition, but that lasted only until we fully realized the car's historical significance, which is good reason for leaving it exactly as it is. With a momentary feeling of nostalgia, I remembered that I last saw it run in 1964.

On the appointed day of the "reunion," we all met at Laguna Seca. Phil and Dan arrived in Phil's Blower Bentley, and the pair of them looked like a couple of retired race drivers who had lost their way to Brooklands. The car was still on the trailer because it was the original trailer, and Mike Shoen wanted to impress us with the authenticity of the whole rig. After being duly impressed, we helped unload the car and Mike gave the keys to Dan, who climbed in.

Mike Shoen pointed out to us that the car still had what was originally known as the "Dan Gurney FIA windshield." When the car was raced, the FIA regulations were very strict in regard to the dimensions of the car. For instance, the car's trunk lid is bulged to accept the regulation FIA suitcase and the windshield was required to be a certain height. Dan and the other drivers were never happy with the windshield and wanted it lower, so a modification was made so it could be inspected in the regulation position, but after the race started the wind blew it down to a position where Dan was comfortable with it.

Dan turned the key and hit the starter button. The engine burst into life with the loud blatting sound characteristic of a tuned V-8. After a minute or two for warming up, he took off up the track like he really meant business and we could hear him accelerating all the way up the hill, then shifting down for the corkscrew and accelerating down the hill before braking hard for the 90-degree turn leading back onto the pit straight. He certainly wasn't hanging around and after several more laps he came in, slammed on the brakes and jumped out with his face wreathed in smiles like he was in his second childhood. His first remark was that it felt like it was only a week ago, instead of 15 years, since he raced the Cobras.

Phil went out next, and he got straight on it too, not to be outdone by Dan. Being able to get Dan and Phil together was a stroke of great good fortune, because the two of them used to spend a lot of time in each other's company and they competed with one another both on and off the track. There are many hilarious stories of the tricks they played on each other.

When Phil came in, he also said the car was exactly as he remembered it and he could recall vividly how it understeered

79

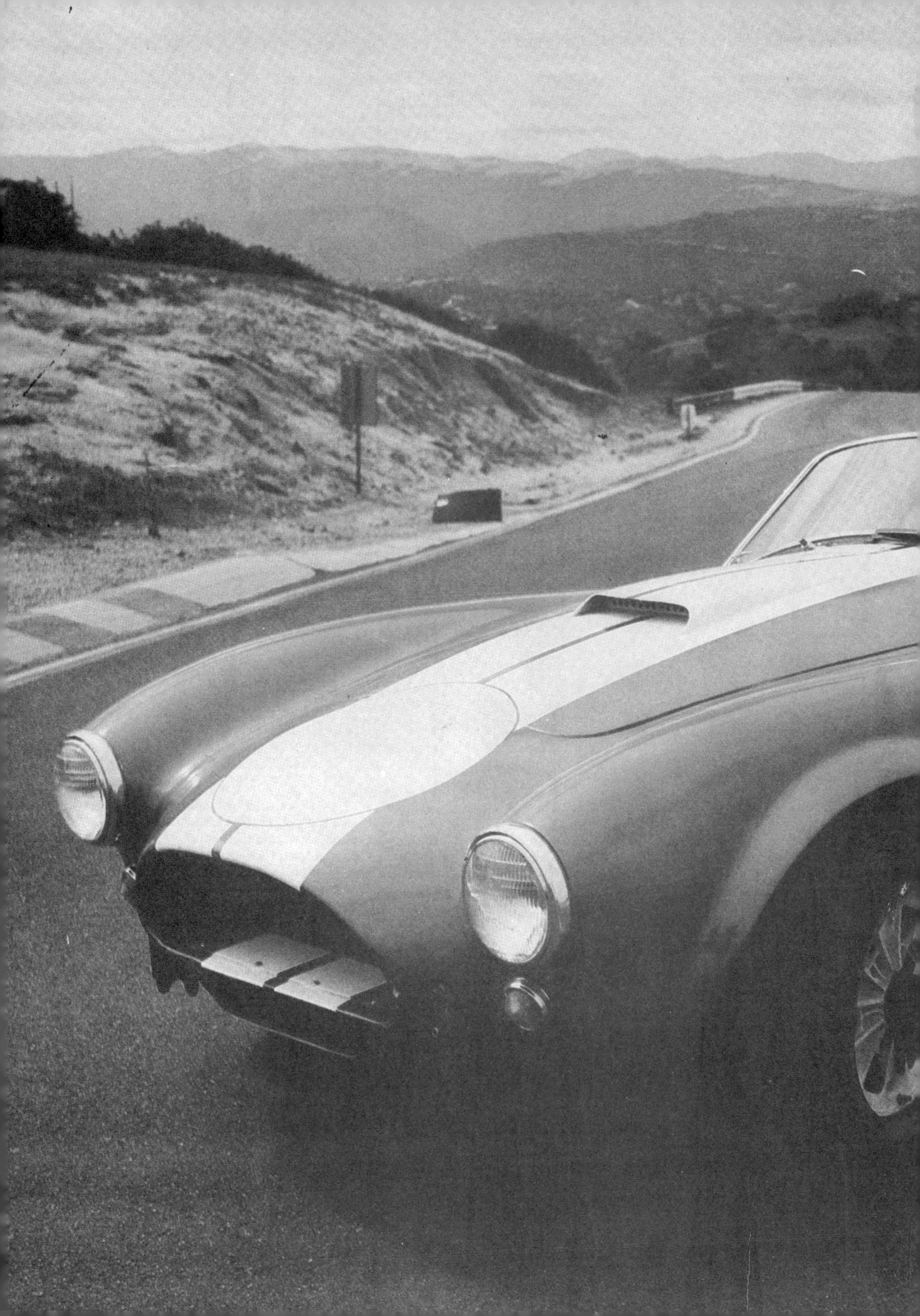

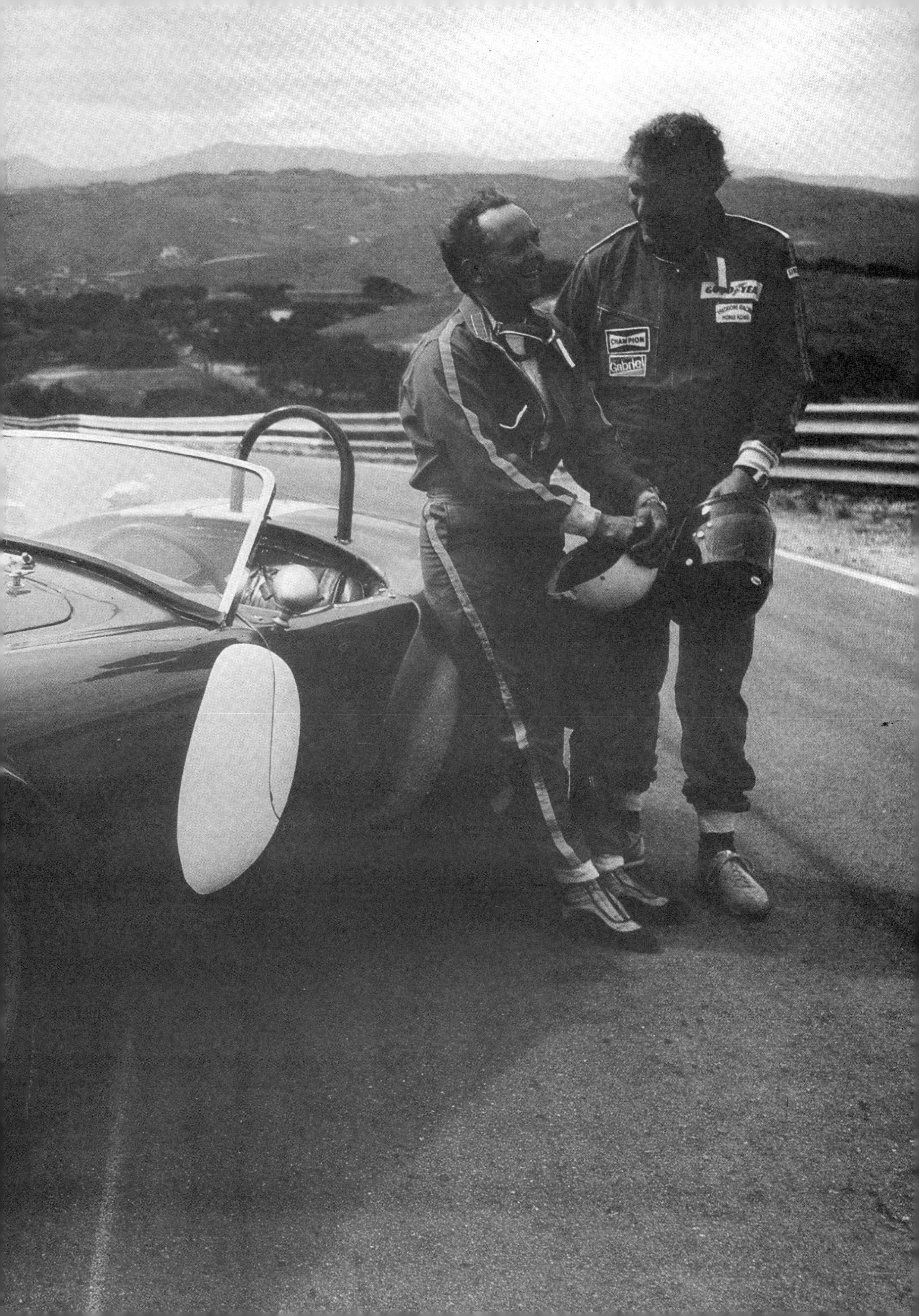
CHAMPION
Gabriel

but that this could easily be corrected with the throttle. He also recalled the considerable amount of brake pad knockoff. This refers to the way in which flexing of the suspension due to hard cornering or a rough surface caused the brake discs to knock the pads back so you had a low pedal when you applied the brakes. Although disconcerting, it was not serious and could be controlled by pressing the brake pedal with the left foot to bring the pads back into position before braking heavily.

To Dan, driving the car brought back a host of happy memories. Driving the Cobras was one of the most exciting times in his life, and the main reason was that it involved a bunch of guys from Los Angeles taking on the cream of the European racing establishment with cars they had designed and built themselves, which was absolutely true.

The man behind it all was Texan Carroll Shelby, who had a long and successful career as a race driver behind him and had only quit because of a heart problem. Before his retirement, Shelby had tried to interest General Motors in a competition sports car, but GM was interested only in Corvettes, so Shelby directed his immense charm and powers of persuasion in Ford's direction, with considerable success. What he actually did was to bring together Ford's engines (and Ford's money) with AC Cars Ltd in England, the builders of the AC Bristol.

In the best tradition of the British specialist car builder, AC Cars is nothing much more than a fabricating shop in the main street of a sleepy little English village called Thames Ditton. At the time, the Bristol Aeroplane Company had just stopped production of the engine used in the AC Bristol and Shelby felt that a small-block Ford V-8 would be ideal. He caused a modified but engineless AC to be flown to Los Angeles, where he and Dean Moon had it running within eight hours. The Cobra was born and Carroll Shelby was on his way.

The true ancestry of the Cobra goes back as far as 1950, when a relatively unknown designer named John Tojeiro decided to offer a racing sports car chassis, which would accept almost any of the available competition engines of the time. Working by himself, Tojeiro built what was eventually to become the basic layout for the Cobra. One of his first customers was named Cliff Davis.

Davis ordered his car with a tuned version of the 6-cylinder Bristol engine. When it came time to body the car, Tojeiro was in a hurry and also somewhat lacking in resources, so he looked around to see what other people were doing. Showing excellent taste, he settled on the beautiful little barchetta body with which Carrozzeria Touring had clothed the Ferrari 166 Mille Miglia, and built a fair copy of it, although he simplified it along the way to suit the available amounts of time and resources. The Tojeiro-Bristol became a famous and successful car and, thanks to Touring, it had an ageless appearance.

At the time, AC Cars was having some bodies built by a company called Buckland Body Works Ltd, which was close to Tojeiro's shop, and Derek Hurlock, who is now the boss of AC, heard about Tojeiro while on a visit to Buckland. A deal was made and Tojeiro cooperated in adapting his design for road use and the car appeared as an AC, first with AC's own 6-cylinder engine and later with the Bristol.

Because it was built in a fabricating shop, the AC Cobra benefited enormously from its extreme simplicity. The chassis consisted of two 3.0-in. diameter steel tubes connected by crossmembers in ladder-style with spring towers at each end. Body framing and support brackets were welded to the main tubes and the bodies were of aluminum. Front and rear suspension was by transverse leaf springs and lower A-arms, and disc brakes were used all around. In the early stages of production, a number of modifications were made from time to time so that there was a certain amount of variation between the cars.

The first contract was for 100 cars, which were delivered between December 1962 and April 1963, and the first 75 of these were fitted with Ford's 260-cu-in. engine. However, Ford was starting to build a 289-cu-in. engine which was absolutely ideal for Shelby's purpose. Ford had been doing extensive research into methods of thin wall casting, so the new engine was light. It had a rather extreme bore to stroke ratio of 4.00 in. to 2.87 in.,

which also helped to reduce the weight as well as keep down the piston speed and permit big valves to be used, and it was offered in a high-performance version giving 271 bhp. It was one of the best engines ever built.

The engine in Mike Shoen's car was, of course, tweaked by Shelby's men, but not excessively so. Shelby's sworn intention was to win with a team of Cobras the series of classic races which counted toward the international manufacturer's championship, or to put it more succinctly in Carroll's words, "To git Ferrari's ass." These races included Le Mans, the Targa Florio, Sebring and other long-distance events, so stamina was just as important as speed. Fortunately, the engines were extremely rugged just as they came out of the box, so the stock crankshafts, rods and pistons were retained after being balanced.

The two engine men were Cecil Bowman and Jack Hoare and their primary job was very careful hand assembly to fine limits, and considerable attention to the gas flow through the heads. Four twin-choke, downdraft Webers were used and the diameter of both the intake and exhaust valves was increased by 1/16 in. To take advantage of the larger valves, the ports were opened up and finally the complete combustion chambers were polished. Various compression ratios were used, but 11.6:1 was about normal. With these and other modifications, such as different valve overlap and ignition timing, the engine put out between 340 and 370 bhp. This output by no means stretched the engines anywhere near the limit, and Phil Hill recalls them as being very reliable as were the whole cars. Furthermore, according to Phil, the torque characteristics of the engines were such that you didn't have to bother very much where you were on the power curve.

When Phil Hill signed up to drive for Shelby, he had just left Ferrari where, although the boss wasn't too easy to get along with, the cars were certainly very strong and reliable. Phil recalls that while he was driving for Ferrari he spent quite a lot of time watching the cars of the various opposing teams just breaking up around the drivers, and he was relieved when the Cobras proved to be strong and reliable too. An amusing incident occurred at Sebring at this time. Phil got into the lead at the start, and when he came around in the Cobra after the 1st lap, the Ferrari mechanics were all standing in front of the pits cheering and waving him on.

Dan Gurney near Cerda, Sicily in a 289 Cobra during the 1964 Targa Florio.

around race cars most of his life and he now works for Dan Gurney, who describes him as "a one-man army." It was probably because of the influence of these two that six competition coupes were built on the 289 FIA chassis. The first coupe was built in Venice to Pete Brock's design and today it is still one of the best looking and cleanest competition coupes ever built. It retained much of the mean look of the roadsters but the windshield was more steeply raked, blending into a long, sloping fastback ending in a chopped off Kammback tail. The cars were called Cobra Daytona Coupes, because the first car made its initial appearance at Daytona in 1964.

One day back in 1964, while they were testing at Riverside International Raceway, Ken Miles took me for a ride in the coupe, which was an interesting experience because not only did it give me an impression of what the coupe itself was like, but also it gave me a good impression of the handling characteristics of the competition Cobras in general, and altogether it was a fairly hairy ride.

For a start, Ken was grinning from ear to ear throughout, because he thought he was scaring the hell out of me (which he was). Then, there was more noise inside the car than out, which meant it was absolutely deafening inside, and also very hot and rather oily. On the back straight we were probably hitting about 150 mph, but it was coming up through the esses that was disconcerting because Ken held the car in a series of classic 4-wheel drifts in which the car was pointing at 45 degrees to its general direction of travel. This, of course, was the classic method of getting a car which underwent considerable camber changes while cornering through a turn as quickly as possible. It wasn't a slow method of cornering, but it was different from what you see today and you had to be strong and courageous to do it.

As far as driving the Cobras in a long-distance race was concerned, Phil and Dan agreed that the cars were very effective but very tough to drive. Phil remembers them as being exceptionally stiffly sprung and much stiffer than the Ferraris he had been driving. On circuits such as the Targa Florio, Phil recalls the cars using up what suspension there was and bottoming all the time, so it was a question of hanging on rather than driving.

Dan's feelings were that you had to be fit to drive them and you had to roll up your sleeves and really go to work. Dan recalled a tense moment at one race during a pit stop. Jerry Grant was about to get in the car and Carroll Shelby was shouting at him to just get in and get going. So Grant did and took off without fastening his seatbelt, but the ride was so wild he was practically being thrown out of the car and had to stop and fasten the belt.

The records show that Mike Shoen's roadster was campaigned as a team car in 1964 at the Targa Florio, Spa and the Nürburgring. It also won the Freiberg Hillclimb and the Sierra Montana Hillclimb in Switzerland in the GT class driven by Bob Bondurant. In 1965, which was the year in which Shelby took the FIA GT championship from Ferrari, it won the Tourist Trophy in the GT class and was 4th overall, driven by Sir John Whitmore, and was then used as a practice car at Spa and the Nürburgring. It finished the season by winning the GT class at the Rossfeld Mountain Hillclimb in Germany driven by Bondurant.

At the end of the season, the car was shipped back to California and sold. It remained in storage for nine years until 1974 when it was taken out, lubricated, tuned, washed and waxed and it has remained that way ever since.

One could describe the Cobras as being relatively unsophisticated, but tough and very fast and they were a source of embarrassment to people with much more sophisticated machinery. When Carroll Shelby set out to campaign his Cobras in the classic European road races, it was an exciting time for everyone and it proved that a bunch of guys from L.A., many of whom had gained their early experience in hot rodding, could take on the sophisticated European racing establishment and beat them at their own game.

Although Carroll Shelby's main competition were the other international sports car teams, he was also competing to some extent for Ford's patronage with Holman & Moody, who were preparing cars for NASCAR events. It was a political situation, and Dan Gurney says he steered well clear of it because, in those days, he was trying to make a career for himself as a driver and didn't want to get involved in politics. Holman & Moody's cars were powered by Ford's 427 competition engine and it was natural Ford should want these engines in the Shelby program.

Presumably because he had learned fairly early on in life that he who pays the piper calls the tune, Carroll and his men did a fast shoehorn job with a 427 engine into a normal Cobra chassis. Ken Miles, who was one of Shelby's employees and a driver of considerable repute, did most of the work and it was Ken who, while practicing for Sebring in 1964, managed to stuff the 427 car into about the only tree growing in that part of Florida. The car was repaired for the race but it suffered a number of ailments before finally retiring on the back of the course.

However, the effort was not in vain because it was the predecessor of a series of some 350 427 Cobras, which were the fastest production sports cars ever built and were radically different under the skin, and considerably more sophisticated than what had gone before. The frame was bigger and stronger, being constructed of 4.0-in. tubes, Texas-size driveline components were used throughout, the suspension used coil springs and upper and lower A-arms front and rear and Halibrand magnesium wheels were standard equipment.

Some competition versions of the 427 were built and I asked Dan and Phil what they were like to drive. Both agreed that they much preferred the 289 cars because the 427 wouldn't really do anything the 289 couldn't do and some things it really didn't seem to do as well. However, they did mention that the 427 might have a slight advantage at a circuit such as Spa, which was tremendously fast with a very smooth surface. Of course, the difference between the street versions of the cars was much less subtle because the stock 289 engine put out an advertised 270 bhp and the 427 about 425 bhp, depending on whom you asked.

Shelby's operation in Venice, California attracted a lot of talented people from the southern California area and among them were Pete Brock and Phil Remington. Remington has been

83

GT40 REBORN

Fortunately, they hadn't thrown away the mold

PHOTOS BY GEOFFREY GODDARD

As Henry N. Manney, III reported more than a decade ago: "Actually the big Fords don't thunder at all ... their silently wuffling progress being a strong contrast to the screaming Ferraris and loudly bellowing Porsches." But thunderous was Ford's impact on the racing world that June of 1966 when three 7.0-liter GT40s outclassed all competitors at Le Mans, sweeping 1st, 2nd and 3rd places.

Mark I, II, III. Broadley. Wyer. Shelby. McLaren and Amon. Miles and Hulme. Gurney and Grant. Was that era really all that long ago?

Now a specialty company in Britain plans to resume GT40 production, so Contributing Editor Doug Nye went to Weybridge, Surrey to check things out. He reports:

The original cars were built by JW Automotive Engineering (nee Ford Advanced Vehicles) in Slough, and that company has retained the design rights through years of inactivity. John Willment and John Wyer were co-founders of JWA and they have now made an arrangement with Peter Thorp's Safir Engineering Limited to construct 25 uprated GT40 Mk Vs. These are not so much replicas as a delayed continuation of the original run, using a slightly modernized design and continuing the original 1000-series chassis numbering system.

I believe 112 official Ford GTs were built during the Sixties as Mark I, II and III cars—not counting the very different factory Mark IVs of 1967. One leading project engineer was Englishman Len Bailey and at the height of the program he worked closely with the British Alan Mann team. Mann's GT40s were very quick indeed. They used modified geometry front suspension with revised bottom wishbone pickups. They often qualified on pole, but always seemed to fail in the race itself. Mann subsequently built the ultimate lightweight GT40 for the wealthy British private entrant Malcolm Guthrie.

Len Bailey has now modernized and redesigned the car to form Safir Engineering's Mark V, using the ultimate Alan Mann spec as his base, and adding a dozen years' or so technical progress. The most notable new features include massive ventilated disc brakes in place of solids and proper CV driveshaft joints replacing the original rubber doughnut-type. The engine is a 289 Ford to customer choice, and the transmission a ZF 5DS25/2 similar to BMW's M1. The predominantly 20- and 22-gauge steel monocoque has been subtly modified to allow Safir's craftsmen to fabricate it throughout, instead of using certain pressed panels as original. The press tools have long since vanished.

Aluminum fuel tanks replace rubber bag type in the tub sills and the radiator is now brass rather than aluminum. The Mark V tub is marginally lighter than the originals while it carries fiberglass body panels taken from JWA's surviving molds. Front suspension uses the Mann-modified geometry, and the adoption of ventilated disc brakes has meant casting new uprights to accommodate them. Whereas the original GT40 front uprights were left- and right-handed, the new Mark V's are one-pattern reversible. New rear uprights have also been found, close to original design.

Bailey recalls that the GT40 program planners preferred simply to choose the most expensive available components rather than those which might prove technically the best. Original-type suspension joints, for example, are no longer available, and Safir uses modern Rose joints throughout, tailored to the job. Original Heim bearings have been discovered in A-1 condition for the inboard lower rear wishbone pivot. Three of Safir's 10 men have contemporary GT40 experience. Two were with JW Automotive and one, Jim Rose who built Thorp's open prototype car virtually single-handed, had been with Alan Mann and with Holman & Moody. Jim recalls that those Heim bearings cost around £60 each in 1966–$134!

Safir's Mark Vs are offered for road or racing use—with road certification being the customer's responsibility—from an initial price of £42,500. In the U.S. that will be a whopping $145,000—without certification. So says Peter Ledwith, General Manager of Wolfshead, the sole U.S. importer (Wolfshead is located at 1547 10th St, Santa Monica, Calif. 90401; 213 395-8484).

The project began when Peter Thorp decided he could not afford the price, nor the aggravation, of an original "£60,000 worth of rusty trouble." He runs a successful airfreight business, and set up Safir in the mid-Seventies as a racing constructor. It ran what had been the Formula 1 Token under the Safir name, and then built F2 and F3 cars. The racing side diminished and Safir began fabricating chassis for the Panther Lima, producing more than 500 before Panther went bankrupt. Safir also designs, develops and constructs 6x4 and 6x6 multi-wheel-drive Range Rover and Land Rover conversions, and has built replica Cobra Mk

II and Mk III chassis. The company developed and built the Rondel Procar Series team's BMW M1s last season and has a fine reputation for Aston Martin service and restoration. The staff totals only 12 but their accumulated experience is immense.

I saw the second and third GT40 Mark V tubs being fabricated, and Thorp's prototype primrose yellow roadster impressed me with its impeccable finish and furnishing. A brief rumble around the old Brooklands service roads and runway confirmed its quality. It feels every inch a well made modern fun car for the wealthy enthusiast—a worthy successor to the GT40 legend. —*Doug Nye*

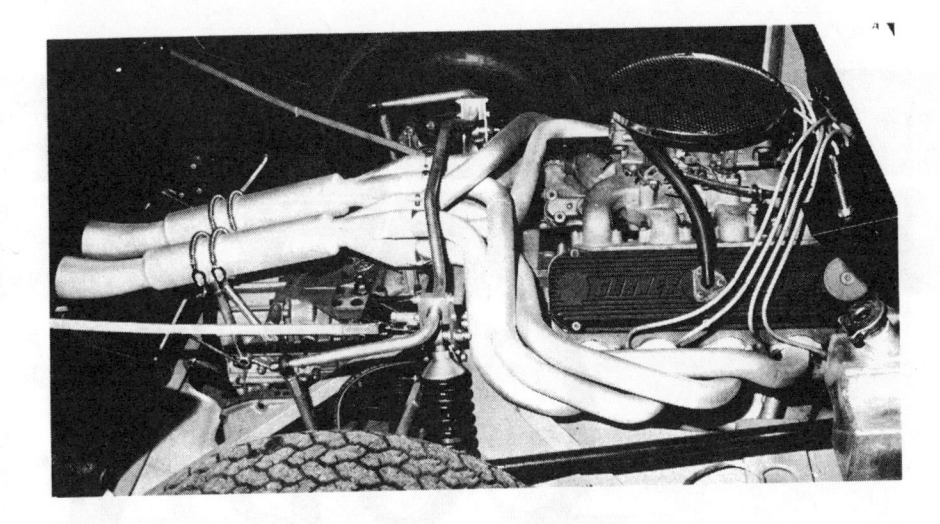

CONTINUED FROM PAGE 17

Ford GT resembles Broadley-designed Lola GT but is aerodynamically cleaner.   *Businesslike driver's compartment.*

FORD GT

including the driver. Cockpit ventilation is always a problem with GT cars, and this has been solved by ducting air from the pressure area under the nose to a position above the instrument panel so that the necessity for opening the side windows is eliminated. At the same time air is ducted to an "air conditioned" driver's seat, which is perforated with a series of holes so that perspiration is evaporated.

Another interesting feature of the seat is that the back is inflated, and this can be adjusted to suit various drivers by a hand pump in the cockpit. Deflation is taken care of by a push button on top of the pump. Another feature designed to improve driver comfort is the pedal layout, which is adjustable in the manner of the original Ford Mustang sports car while the seat is a permanent part of the main structure.

The car was designed to take the Ford 4.2-liter Indianapolis engine, which develops 350 bhp, but the eventual power unit is likely to be the Ford twin-cam Indy engine. The engine is mounted in the middle behind the driver. The drive is taken to the transmission by an English Borg & Beck multi-plate clutch of 7.5-in. diameter, and the transmission itself it a Type 37 Colotti. This transmission has the disadvantage of only four speeds, but at the time it was the only proven transmission available which would withstand the torque of the engine.

The front suspension is conventional with upper and lower A-arms, and the rear suspension uses lower inverted A-arms with transverse links at the top and long trailing links from the main bulkhead. The inboard universal joints on the axles are Metalastic units which eliminate the need for splines. The Girling disc brakes are mounted inboard both front and rear and 11.5-in. discs are used, cooled by air ducted from the front intake and two additional intakes, one on each side of the body.

It is difficult to predict how the Ford GT will perform under actual racing conditions, but it is evident that is is an extremely well-engineered car constructed with all the resources of the Ford Motor Company and using some of the best brains in the racing business. GT racing is a fiercely competitive branch of the sport, with an enormous spectator following in Europe, and, in consequence, the effect on sales of a win is considerable. Obviously, this is what Ford has in mind.

However, when all is said and done, Ford must be complimented for having the courage to put its reputation on the line against the full might of Ferrari, and it will certainly be interesting to see the outcome of its efforts during the 1964 season.

"Spoiler" with lights and scoops improved aerodynamics.

COBRA REPLICAS

The ultimate do-it-yourself performance kits

BY TONY HOGG

IT APPEARS THAT the decade of the Eighties is turning into the decade of the replicas because one sees them on the streets in every shape, size and form and they are widely advertised in kits, partially assembled kits or as complete cars. Actually, this has been going on for a long time, but only recently has it really blossomed. Among the many replicas available are about eight versions of the Cobra, so we rented Riverside International Raceway for a day and assembled two of the better known and established replicas plus an original to use as a benchmark. We also asked Carroll Shelby, who after all was the instigator of the whole thing, to come along to comment, which he did.

The two replicas we selected were the ERA made by ERA Replica Automobiles of New Britain, Connecticut and the Contemporary Cobra Replica made by Contemporary Classic Motor Car Co Inc, Mount Vernon, New York. The reason for selecting these two is that both companies have sold quite a number of cars and the cars themselves seem to be well engineered.

The trouble with the replica business is that it tends to attract a variety of people. Some are on an ego trip, some are sincere, honest and well intentioned, but many of them have no conception of the amount of financing involved. And then there are a few who have a roguish trend, but there is always a bad apple in every barrel. It is for these reasons that we at *Road & Track* have been somewhat cautious in our coverage of replicas.

Many of the kit cars and replicas of all kinds on the road never cease to amaze me, not only because some of them are incredibly ugly, but also because they are complicated and expensive to build. In the case of the fiberglass Cobra replicas, this is not true because the body is relatively easy to produce and it is one of those timeless designs that will always look exciting.

The reason for the relative simplicity of the Cobra is that Carroll Shelby was underfinanced when he started the project. At the beginning of the Sixties he was a successful race driver but he had a bad heart condition that precluded any more racing. Bringing his entrepreneurial instincts to bear, he went to AC Cars Ltd in England, which was then building a rather handsome sports car on a simple tubular frame using transverse leaf springs at each end. Shelby felt that this car would accept Ford's 260-cu-in. engine and be a marketable proposition. Prototypes were built and an order for 100 cars was placed, which were delivered early in 1963. Actually, only the first 75 cars had the 260-cu-in. engine because, at that time, Ford introduced the 289-cu-in. model that was offered in a high performance version putting out 271 bhp.

By now Ford was very interested in Shelby's project and the resulting racing successes. So, it was decided to redesign the car to take Ford's 427-cu-in. engine. Keeping things simple, the redesigning consisted mainly of using larger-diameter frame tubing and modernizing the suspension by using unequal-length A-arms with coil springs front and rear. The result was absolutely stunning in both performance and appearance. To accommodate the additional machinery, the car had to put on some fat so it is a much chunkier-looking car than the earlier 289; however, it is no more difficult to reproduce in fiberglass. Actually, it's a most interesting body because it was copied unashamedly in 1950 by English designer John Tojeiro from a Touring-bodied Ferrari 166 Mille Miglia.

For these reasons, the 427 Cobra is a relatively simple car to

PHOTOS BY JOHN LAMM

Replica by Contemporary Classic.

Original Cobra.

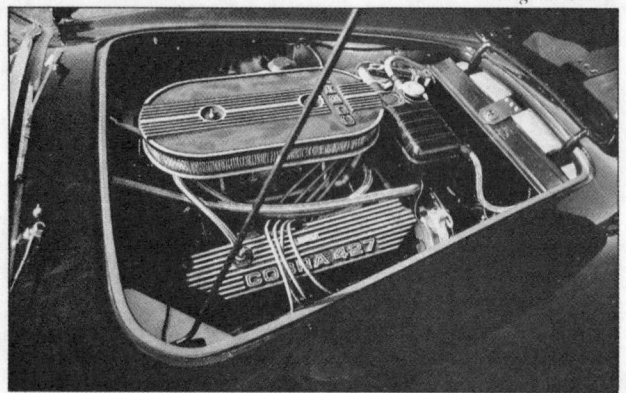

Original Cobra.

Replica by ERA.

Replica by ERA.

Replica by Contemporary Classic.

Replica by Contemporary Classic.

88

build today and its uncluttered fiberglass body doesn't present too many problems to anyone who has worked around boatyards. Furthermore, the car's mechanical components are not too hard to find and the knowledge, skill and experience required are a lot less than those required for building, say, a replica of a Type 35 Bugatti. In fact, it is virtually impossible to tell a replica Cobra from an original.

When we arrived at Riverside Raceway, we found one or two people had brought their buddies along so we ended up with five cars, including one 289, and it was interesting to learn that four of them had been driven considerable distances to the track. The exception was the Contemporary which, having set wheel to ground only a few days before, was a bit of an unknown quantity and consequently arrived on a trailer. Unfortunately, it was also lacking a hood because the height of the intake manifold precluded the use of the normal panel and there had not been time to fabricate another one.

When I first asked Shelby what his feelings were about people building replicas of Cobras, he replied that he couldn't understand why the hell anyone would want to build a replica of a car that was already 20 years old when he built it. An interesting observation, but I suspect he is flattered and, although he doesn't knock replicas in any way, he does doubt the financial situation of some of the manufacturers. Once we had gotten ourselves all fired up and ready to go, the answer to my question to Shelby about replicas became obvious if for no other reason than that the cars were blindingly fast but handled, stopped, steered and behaved generally very well provided one paid due respect to the available power. Having a 427 myself, I recall writing a piece about Cobras for R&T in the July 1974 issue in which I said, "The Cobra is nothing more than a weapon designed specifically for proceeding from one point to another in the minimum amount of time."

Getting maximum-speed times for a very fast car at Riverside is difficult because the main straight has a sharp dogleg which can get a bit scary, and then there is the problem of asking the owner of a fast and extremely valuable car to let you belt the hell out of it. In addition, getting the most out of cars such as these requires a little practice.

The original 427, which was very kindly provided by Ron McClure, proved to be the fastest. It did 0–60 mph in 4.8 seconds and the ¼-mile in 13.3 sec with a terminal speed of 107.0 mph. The ERA, which was the factory demonstrator, did 0–60 mph in 5.6 sec and the quarter in 13.9 at 101.7 mph. From 0–100 mph, the original car took 11.3 sec compared to the ERA's 13.2 sec. Unfortunately, we were unable to get reliable figures on the Contemporary because it developed a bad wheel balance problem at higher speeds.

However, some months ago we had been able to drive the Contemporary demonstrator car and got some figures at a different location and under different conditions. That car did the quarter in about 13.5 sec at approximately 105.5 mph. These figures compare favorably with the figures we published in July 1974 for my own car, which showed the quarter at 13.8 sec with a terminal speed of 106.0 mph. Incidentally the 289 Cobra we tested in June 1964 was no slouch either because it did the quarter-mile in 14.0 sec with a terminal speed of 99.5 mph.

The major difference between the Contemporary and the ERA is that the ERA's frame tubes are rectangular but the Contemporary has round frame tubes as had the original cars. Both replicas use Jaguar independent suspension in the rear and the Contemporary uses E-Type Jag suspension in front. Because the rear suspension was used on the sedans as well as the sports cars, it is a relatively easy unit to obtain. Unfortunately, the E-Type front suspension was unique to that model and therefore rarer and more expensive. ERA, on the other hand, uses current Ford independent front suspension components.

With regard to the engines, there is a certain amount of confusion. The original 427s came equipped with racing 427 engines that were side-oilers and had cross-bolted main bearings. However, the supply of true 427s was limited and 428s were sometimes used instead. In fact, the two engines were used somewhat randomly so one batch of cars would have 427s and another batch might have 428s. The 428 was a totally different engine and was used to power the bigger models in the Ford and Lincoln-Mercury lines. However, as Carroll Shelby says, how do you tell the difference between 425 bhp and 475 bhp? Comparing the two engines, Shelby says that the 428 is a bit quicker at the bottom end of the range but the 427 starts to build up at the top end. In addition, Shelby told me that the 427 would occasionally throw a rod for no particular reason when you were cruising along at 3000 rpm. Fortunately, I'm glad to say, this has not been my experience.

Obtaining a suitable transmission to handle the power of a 427 or 428 engine is no particular problem because plenty of the original Ford top-loaders are available, and it's a beautiful transmission, well able to handle the highest torque outputs. Alternatively, you can get a Doug Nash 5-speed, although the gears in the original gearbox are so well spaced for such a torquey engine that five are not really necessary. For those people who don't like to shift, Ford's C6 automatic does the job nicely.

One of the greatest advantges of buying a replica, particularly if you have sufficient mechanical skill to buy it in kit form and assemble it yourself, is that you can tailor the car to your own particular desires and requirements. For instance, the original 427 engines came with a compression ratio of 11.5:1, but that was in the days when you could buy gasoline that was in excess of 100 octane at almost any gas station. Today, that kind of compression ratio would require either avgas or a big and expensive dose of some sort of additive. Therefore, the answer is to go to a much lower ratio. Actually, the power-to-weight ratio of a Cobra is such that almost any V-8, used or new, would give the car pretty sensational performance.

Just because you are buying a Cobra replica doesn't mean that you have to have a 427 or 428 engine. The Ford 351 Cleveland engine makes a good power unit for the car and can be obtained in various stages of tune. One reason for not electing to use a true 427 is that it will probably cost you a good $3000 from a reputable engine rebuilder. In addition, there are many Chevrolet enthusiasts around and quite a few of the Cobra replicas are Chevy-powered.

ERA and Contemporary have somewhat different methods of marketing their products. ERA prefers to sell one kit for $14,800 that is a bolt-together design with no fabricating at all. What you get is a complete body and chassis with all instruments, cooling system, fuel system, electrical system, windshield, bumpers, etc, and these are either already installed or ready for installation when the running gear has been obtained. Also included in the kit is a list of part numbers and sources for various minor items such as radiator hoses, fan belts and other pieces that will depend on the type of power unit used. What you have to provide are the engine, transmission, Jaguar final drive/rear suspension unit, Datsun Z-car rack and pinion, Triumph Spitfire steering column, Chevrolet Camaro disc brakes, front coil over shock units, steering wheel, road wheels and tires and MGB windshield wiper motor.

On the other hand, Contemporary offers three different kits ranging from a body/frame unit at $6750 to a package similar to ERA's at $13,450 and, in between, a basic home-builder package at $8150. Also, any of the three packages can be had with totally rebuilt and assembled suspension and brakes for an additional $4250. Depending on the depth of your pocket and your available time and skill, you could complete a Cobra replica for about $20,000, but for those who are busy and not so adept, the best way is to buy the most comprehensive kit.

Both ERA and Contemporary claim that virtually all the kits they have sold have either been completed or are being worked on actively. This is because most of the customers are true enthusiasts who have always wanted something like a Cobra but have never been able to afford one.

Our time spent at Riverside was unfortunately all too short. Carroll Shelby enjoyed himself immensely, perhaps because although he owns one of each of the various Cobra models, he doesn't drive them much, preferring to dash around the back roads of Texas on a fast motorcycle. His general opinion of the replicas was that they were just as quick as the originals but they had softer suspensions; for really fast driving over varying conditions, an original would be a bit quicker. On the other hand, he felt the replicas were ideal for everyday road driving.

What we learned at Riverside, and I knew from owning one, is that you have to treat a Cobra with respect and it doesn't matter whether it's an original or a replica. With a weight of something over 2500 lb and perhaps 450 bhp available, depending on the engine, you have a power-to-weight ratio that is something fierce.

Perhaps one can sum up the 427 Cobra by saying that the late Ken Miles, who was a Cobra driver and developer, once took one of the cars from 0–100 mph and back down to a complete stop in 13.8 sec. With $20,000 and some knowledge and skill, perhaps you could go out and do likewise.

THE CANADIAN COBRA

TOTALLY DIFFERENT IN concept to the ERA and Contemporary Classic Cobras is the Aurora made by Aurora cars of Richmond Hill, Ontario, Canada. This replica can only be bought as a complete car. However, it is equally desirable to the less mechanically adept enthusiast. Furthermore, it is a replica of the earlier 289 version of the Cobra, which some people prefer to the more chunky 427.

It meets all emissions and safety regulations and carries a warranty of six months or 10,000 miles. Obviously it caters to a totally different market and when I talked to the Aurora folks they said their average buyer was between 35 and 50 and was fulfilling his dream of owning a sports car totally different from anything else on his block. Put in its proper context, the car meets these requirements very well indeed. What it is, in fact, is a modern high performance sports car clothed with a Cobra body, and there is absolutely nothing wrong in that.

When Carroll Shelby first went to work on the 289 Cobra, it was even then somewhat archaic. To attempt to re-create the 289 two decades later would obviously be a mistake and so there is virtually nothing in the Aurora that even resembles the 289 except the body, which is an exact replica.

Aurora builds its own tubular frame and then relies on Ford for some of the components. The engine is the stock Ford 302-cu-in. V-8 and it is mated to either Ford's 4-speed manual transmission or to an automatic. The suspension is fully independent and the brakes are 10.0-in. solid discs mounted inboard at the rear and 9.5-in. ventilated discs in front. Steering is by rack and pinion and the standard tires are Michelin TRX with Pirelli Cinturatos as an option. The finish is very good, with the fiberglass panels being made by C&C Yachts, which has an excellent reputation in the yachting world, and the interior is done in real leather. Unfortunately, the best things in life are not free and the Aurora comes out at $36,000.

On the road, the car inspired confidence, particularly in braking and steering, and we couldn't quibble about the roadholding either. Out at the track, we managed to get a ¼-mile time of 14.9 sec and a terminal speed of 91.0 mph, while 0–100 mph came up in 18.8 sec.

Although these times and speeds are no where near those of the 427 Cobras, they are still impressive. With this kind of performance in a well designed car using many stock components, mostly of Ford origin, the Aurora provides a good alternative for someone who wants a Cobra replica but lacks either the time or the mechanical knowledge to assemble a kit. —*Tony Hogg*

Profile:
CARROLL SHELBY
Wily competitor

BY TONY HOGG

THE FIRST TIME I met Carroll Shelby was at Sebring in 1964 when he was managing the Cobra team. On that occasion he was confined to a wheelchair because of a leg injury and all the time muttering something about "gittin' Ferrari's ass." Whether or not he achieved his ambition in that direction is a matter of history. The last time I saw him was recently in his office and later when we were both judging a chili cook-off in Dallas, and he brought me up to date on what had been going on since our first meeting.

Actually, I had seen Shelby on many previous occasions behind the wheel of some really vicious machinery such as 4.9-liter Ferraris, which were very long on acceleration and maximum speed, but a bit short when it came to handling and braking.

Carroll came out of Leesburg, Texas where he was born in 1923, the son of a mail carrier. The advantage to being born the son of a mail carrier in those days was that, by the time you were old enough to know what was going on, deliveries were being made by automobiles instead of by horses and buggies. That was an enormous advantage to any young man's education if a career in automobiles happened to be his particular ambition. Carroll's education came on strong when his father was transferred to Dallas where Carroll was able to watch and help out at the local dirt track ovals. After leaving school, he was variously an oilfield roughneck, home builder, dump truck operator, flying instructor in the Army Air Corps and bankrupt chicken farmer, all of which must have been an interesting and instructive beginning to what has turned out to be a most active and profitable life.

Carroll started racing in an MG TC and, using the practice followed by most successful but impecunious race drivers, which is OPM (other people's money), the MG was borrowed. Unfortunately, you can't make much of a name for yourself as a race driver with an MG TC, and Shelby is most grateful to Luigi Chinetti for getting him some rides in Ferraris, which established him as a professional and sought-after race driver. Unfortunately, in those days the Sports Car Club of America had a firm hold on road racing and the club considered racing for money vulgar and unseemly. On the other hand, there were a few wealthy patrons in the Southwest who owned cars but hired other people to drive them, primarily for nothing more than expenses. One such owner was Roy Cherryhomes, who owned a Cadillac Allard. With this car, Carroll established himself and went on to greater things, with the help of such benefactors as Tony Paravano and John Edgar. He became SCCA national champion in 1956 and 1957, among other successes.

While all this was going on, Shelby had caught the eye of John Wyer, Aston Martin's team manager, and Wyer hired him as a team driver in 1954. This proved to be a most successful association and it culminated in a win at Le Mans for Shelby and partner Roy Salvadori. Meanwhile, Shelby was racing in other events, up to and including the Grand Prix level, first in rather worn Maseratis for the independent Centro Sud team and then in 1959 for Aston Martin, whose cars were beautiful but too long in arriving on the scene to be competitive. Not only was Shelby a reliable and successful driver, but also he was an excellent public relations man for any team he drove for because of his easygoing Texas drawl, cowboy hat and the bib overalls that became his trademark.

However, around 1960 things were not going well for Carroll in other directions because he was starting to suffer severe pain in the area of his heart and this ultimately led to two bypass operations, although, as he says, he never pays any attention to his heart condition and leads a very busy life. But, during the last two years of his racing career he was popping nitroglycerin pills to withstand the mental and physical strain.

Actually, behind Carroll's easygoing charm lurks a very astute mind, and when I asked him why so many ex-race drivers just disappear into the woodwork while he is a

PHOTO BY DOROTHY CLENDENIN

91

SHELBY

very successful businessman, he said that most drivers give no thought to the future, but he had already decided to quit racing and pursue other activities even before his heart started acting up.

The particular activity Shelby had in mind when he quit driving was to build cars and form a racing team—nothing, of course, being impossible to Carroll Shelby. In fact, he will tell you that he really just used racing, which was not much of a money-making proposition in those days, to establish his name and capabilities for his next career, which was, of course, the formation of the Cobra team on the OPM system using Ford's money.

What really started the whole Cobra affair was that Shelby had heard over the grapevine that Bristol Aeroplane Company in England was going out of the automobile engine building business, and this left AC Cars in England without a suitable power unit. Shelby got off an airmail letter to Derek Hurlock of AC, suggesting that one of the smaller American V-8s might just do the job, having in mind the Buick aluminum engine. Meanwhile, in his capacity as a representative of Goodyear, he got to know Dave Evans of Ford and from Dave he found out about the very lightweight iron cylinder blocks that Ford was casting, which did the job just as well as aluminum but without all the complication. This engine had a capacity of 260 cu in. and it was mated to a transmission that would withstand 400 lb-ft of torque.

The original prototype quickly took shape at AC in England and in Los Angeles, and it had the blessing of Don Frey, who was in charge of engineering at the Ford Division. Shelby lost no time in tapping the vast pool of talented people in the Los Angeles area and engaged the help of such people as Ray Brock and Phil Remington. The result was that by the end of 1962 some 75 cars had been sold, which wasn't bad considering production had only started in March of that year. Other important points were that the car was homologated as a Grand Touring automobile and plans for an extensive racing program were mapped out, while the 289-cu-in. engine was used to replace the 260.

The beautiful Cobra competition coupes were brought into action and were quickly followed by the red-blooded 427 cars. At the time, Ford's racing effort was enormous and all-encompassing, both in the U.S. and Europe, and Shelby even became involved in the Mustang program with the Shelby GT350s and GT500s.

Meanwhile, Shelby was forced into competition with Holman & Moody, who were responsible for a part of Ford's racing effort, and the two teams were competing in the same events. The cars were the GT40s, which replaced the aging Cobras and, although some memorable battles were fought, Shelby's control and interest started to fade to a point where he was supervising Trans-Am events but not doing too much else. Finally in 1970, he concluded his association with Ford.

How and why Carroll Shelby ended his association with Ford is not really very clear. Obviously, there was an enormous amount of money floating around, creating considerable temptation to certain people. However, he had been paid a handsome salary for his work and was ready to move on.

To this day he is not interested in becoming involved in racing to any great extent because he feels that racing is an absolutely full-time commitment. And, apart from having many other interests, he feels that he is now too old to make that kind of investment in time and energy. Shelby also feels the whole business of racing has become far too sophisticated.

When he left Ford, he already had a Goodyear tire franchise in the Los Angeles area and he is now heavily involved with custom wheels, mainly for smaller cars because that's where he feels the action is. He also has the famous chili business, a food packaging business, an embryo cattle transplant business, a penchant for buying and selling raw land and a gold mine in Africa.

All of this activity demands a considerable amount of travel and he has his own plane and pilot for the shorter trips. It's apparent Shelby leads a restless existence that suits him well, and he is constantly on the move. He was married when quite young and has three children of whom he is very proud, but he was divorced many years ago and now describes himself as not being marriage material. However, he says he has a wonderfully understanding girlfriend.

Going to his office in Los Angeles was like old home week because the first person I met was Al Dowd, who was the Cobra team manager. Then there is Lew Spencer, who was a driver and is now Shelby's personal assistant.

Carroll Shelby has led an exciting but very turbulent life and seems now to have settled into more of a routine, although still filled with wanderlust. It is apparent that his new association with Chrysler is critical to him because the next peak he wants to climb is to be successful at building low-cost sports cars. Then, he intends to bow out and breed miniature horses and ride his motorcycles around the back roads of Texas.